Spacesuit

Spacesuit FASHIONING APOLLO

Nicholas
de Monchaux

THE MIT PRESS
CAMBRIDGE, MASSACHUSETTS
LONDON, ENGLAND

MIT Press books may be purchased at special quantity
discounts for business or sales promotional use. For
information, please email special_sales@mitpress
.mit.edu or write to Special Sales Department, The MIT
Press, 55 Hayward Street, Cambridge, MA 02142.

This book was set in Helvetica Neue Pro by The MIT Press.
Printed and bound in Spain.

Library of Congress Cataloging-in-Publication Data
De Monchaux, Nicholas, 1973–
Spacesuit : fashioning Apollo / Nicholas
de Monchaux.
     p. cm.
 Includes bibliographical references and index.
 ISBN 978-0-262-01520-2 (hardcover : alk. paper)
1. Space suits—United States. 2. Manned space flight—
United States—History. 3. Technology—Social aspects—
United States. I. Title.
 TL1550.D46 2011
 629.47′72—dc22
                                        2010022786

10 9 8 7 6 5 4 3 2 1

TO KATE AND CHARLES

Contents

Foreword

To float, free from the earth—in "space"—would appear to be the most anti-architectural situation one could imagine. The spacesuit is an otherworldly architecture designed to mediate what is literally uninhabitable. And yet the liberation of the body from gravity points to the many kinds of freedoms with which the most contemporary architecture is invested. As a practical matter, we design for a world that has yet to come; because of the accelerated nature of change, precedents are decreasingly useful; we have to draw from models that engage the unfamiliar.

While we imagine the freedom of the spacewalk as one fully unencumbered, the experience of the astronaut barely corresponds. In order to free oneself of gravity, one has to inhabit a cumbersome prosthetic, complete with its own complex internal biological system (including controls for water, air, etc.); one becomes acutely aware of the fragility of one's freedom. When we examine the history of the spacesuit, we find a discourse that migrated from protection and stiffness toward softness and contamination. Instead of a controlled hybrid of man and machine, an engineered cyborg, we are treated to a more intimate, secret history; of a spacesuit made by a bra company on Singer machines, of astronauts' skin impregnated with silicon to protect them from their own bodily excretions, and of the violent, if secret, rejection of technological monitoring by the astronauts of Apollo 13 and Skylab 3. While critics have compared our studio's work to the most masculine of these precedents (for instance Michael Sorkin and Georges Teyssot's discussions of our work and the AX-3 and AX-5 hard spacesuit prototypes), the current study elucidates a subtler, softer relationship to technology and intimacy than the surface appearance—and one that may be necessary to negotiate our current architectural condition.

Within this multilayered history, we are not afforded distinctions between absolutes, rather the confusion and elimination of them. Within the controlled environment of the space race, the body proves to be an uncontrolled variable, precisely out of control. Our bodies, and the larger ecology, are much less controllable than our ideas of them. Technology, traditionally a tool for ordering the world, inevitably is unable to dominate nature. What is left after we understand this failure?

The juxtapositions contained within Nicholas de Monchaux's narrative describe the practical irrationalities of a design process that ventures into the unknown; Dior and the defense establishment, Cronkite and cyborgs, Kubrick and Kennedy all contribute unanticipated components to a solution whose terms are repeatedly and necessarily inverted. As a designer and writer, de Monchaux is able to provide a perspective that helps us see technology as a collaborative, nonlinear cultural phenomenon rather than as an abstract domain of exclusive expertise and inevitable progress. In contemporary architectural culture, the reflection on technology and nature contributed here represents a potentially instrumental rereading of technology's latest utopian promises. A critical understanding of our own inability to control the world, it turns out, is essential to shaping it.

Elizabeth Diller

Acknowledgments

The history of this book begins with a paper written for Georges Teyssot at the Princeton University School of Architecture in 1998, for a seminar on domestic space. I am particularly grateful to Georges for encouraging my interplanetary interests at the time, and for his continued interest in the work.

Credit for reviving the paper goes to the Thomas Jefferson Society of Architectural Historians and John Maciuika for their invitation to give a lecture in 2002; the spacesuit was the topic of my own expertise that, we decided, best suited the spirit and history of Jefferson: architect, gadgeteer, and captain of exploration. The support and encouragement of my colleagues at the University of Virginia, especially Judith Kinnard and Elizabeth K. Meyer (who came to hear the lecture in a hailstorm!), provided the first clue that the history of a single object could expand to encompass much of my larger interests in technology, nature, and their mutual intersection. Yet another valued University of Virginia colleague, Tiha von Ghyczy of the Darden School of Business, provided essential insight on the relationship between these organizational ideas and the traditional world of agencies, firms, and institutions.

A final, essential catalyst to this book was the generosity of Walter Fontana, who invited me to present the Apollo spacesuit's story as a public lecture of the Santa Fe Institute in 2003. Beginning with an extended period of residency in the spring of that year, the Santa Fe Institute has provided a generous intellectual and creative home for much of the project's imagining and execution. Over the following years, I have been particularly grateful to many others at the institute, including Luis Bettencourt, Jim Crutchfield, David Krakauer, and Andreas Wagner, for sharing essential insights on the nature of human environments from the intimate to the global, and engaging the relevance of these insights to the world of design.

For all this project's ambition, the first person to suggest that it could and should be expanded into an entire book was Kevin Lippert. He has a lot to answer for, and to him I am particularly grateful for his ongoing support and advice. Further encouragement came from Miko McGinty, whose expertise on the world of publishing beautiful things has been—frequently—essential.

Once the book project was launched, crucial mission support came from the University of Virginia and from the MacDowell Colony, which afforded me an extended residency in which much of the larger work's outline took shape. A 2003 grant from the Graham Foundation for Advanced Studies in the Fine Arts provided essential funding for early interviews and archival research.

I am especially grateful for an early interview granted by Homer Rheim, senior manager of ILC's Apollo contributions and later CEO of ILC Dover, Inc. For further guidance in engaging and extending the history of the Apollo suit, I owe a particular debt to William Ayrey, who serves as the company's own historian in addition to his role in the Suit Testing Laboratory for Shuttle and ISS-era pressure suits. Bill has given generously of his own archives and expertise, and with his help I was able to locate and interview a remarkable cast of Apollo-era characters, including seamstresses Eleanor Foraker and Roberta Pilkenton, and managers John McMullen, Thomas Pribanic, and Richard McGahey. I am grateful also to Ken Thomas of the Hamilton Standard Corporation, who provided vital documentation of the firm's interlinked history with ILC, past and present.

For all of this, however, this book would be decidedly thinner and undernourished if not for the support of a research fellowship in 2005–2006 from the National Air and Space Museum of the Smithsonian Institution. After several years of puzzling through Apollo from a perch outside the nation's greatest archive-cum-reliquary to the space race, the chance to see, touch, and engage the nation's spacesuit collection, and share offices with its distinguished body of experts on the space race, was the sine qua non of this book's robust completion. I shall forever treasure the particular vertigo that comes from looking up at the moon after a day of handling objects that lived and worked there.

At the Smithsonian, individual thanks go to Amanda Young, longtime guardian of the suits' physical welfare, as well as to those in the Space History office whose guidance and expertise provided essential detail and insight. They include Paul Ceruzzi, Tom Crouch, Roger Launius, Cathy Lewis, Allan Needell, and Margaret Weitekamp. Fellows with me at the museum, and valued interlocutors on their own areas of expertise, were James Fleming and Dennis Jenkins, whose company I enjoyed enormously. Doug Erwin of the National Museum of Natural History provided additional insight on the larger questions of form and evolution that are central to the book.

With the Smithsonian's assistance, I was able to encounter and interview participants in the drama of Apollo whose firsthand knowledge proved enormously invaluable. These include first and foremost Apollo astronauts Thomas P. Stafford and Alan Bean as well as NASA engineer Joe Kosmo, who still plays a central role in NASA's Crew Equipment Division.

Fulfilling this book's cross-disciplinary ambitions, further essential firsthand accounts were provided by Frederick Ordway III (assistant to both Wernher von Braun and Stanley Kubrick) and Harold Finger (a former administrator of both Apollo-era NASA and Nixon-era HUD). In a more literally cerebral context, I am grateful to Stuart Moss, Terry O'Keefe, and Tom Cooper of the Nathan Kline Institute at the Rockland State Hospital for their memories of Dr. Kline and psychiatry's cybernetic era, as well as, of course, to Manfred Clynes, who provided his own essential recollections.

I owe a great debt to two engineers of "hard" spacesuits who graciously opened their (adjoining!) homes and archives. Hubert "Vic" Vykukal and William Elkins could not have been more generous in their provision of anecdote, insight, and archival material.

More traditional archives providing vital contributions to the book include the Still Picture and Photography Collection of the National Archives, College Park, Maryland, without which this would be a much less beautiful story to behold. Essential documentary material was provided by the Johnson Space Center Archive at the University of Houston-Clear Lake, the National Archives at Fort Worth, Texas, the Archives of the John F. Kennedy Presidential Library, the Archives of the

Lyndon B. Johnson Library, the History Office of the National Aeronautics and Space Agency, the Charles Babbage Collection of the University of Minnesota, and the archives of Davis Brody Bond in New York (where I owe Carl Krebs particular thanks). I am grateful to photographer Michael Light, who provided great advice on suit-centric NASA photography from the Apollo era.

Since 2006, I have been on the faculty of Architecture and Urban Design at the University of California, Berkeley, where the work of shaping the final manuscript took place. Here, I am grateful to several research assistants, including Clare Robinson and Christian Cutul, who helped enormously in finalizing footnotes and references. (Their work joins the essential contribution of Oren Abeles, who first tracked references for me in 2002.) In the Library of the College of Environmental Design, I am grateful to both Elizabeth Byrne and David Eifler for their support of my research, and especially to Matthew Prutsman for procuring a range of obscure books, documents, and reports from UC's furthest reaches.

I thank in particular my supportive colleagues here in California for their help and support in bringing this manuscript to completion. Special thanks go to Anthony Burke and Andrew Shanken, who each reviewed chapters and portions of the work, and to Lucy Jacobs, whose insights on her family's history in nature and urbanism was detailed and invaluable. Further afield, I am grateful for the affirmation from Cynthia Davidson of the story's particular relevance to contemporary architectural practice.

This book's trajectory was finally brought to a successful splashdown with the support and encouragement of Roger Conover and the MIT Press. I am grateful especially to Anar Badalov, Matthew Abbate, Gillian Beaumont, and Margarita Encomienda at the Press for their insight and labor in the book's production.

Particular thanks go to Ceara O'Leary, my incredibly valued assistant in the manuscript's completion. Without her copy-editing and organizational skills, this book would still reside only in virtual space.

And finally, thanks to my own family: first, to my brother, and longtime spare brain, Thomas de Monchaux, without whose encouragement and insight this book would not exist. The lessons of my father and mother, John and Suzanne de Monchaux, have ranged from the robust messiness of cities to the grammatical precision of language. They have enriched this and many other fruits of my own professional life.

My most devoted debt of gratitude goes to my wife and partner in all adventures, Kathryn Moll. She has not known me without this work in some form of its incompleteness, and her insight, encouragement, and love throughout its steady conclusion have been my salvation. And while our son Charles's new life spans only this book's latter stages, his humor and curiosity have been a welcome and wonderful reminder of what it is to delight in the world.

My thanks to all.

Berkeley, California, 2010

Layer **1** INTRODUCTION

Astronaut Edwin E. "Buzz" Aldrin on the surface of the moon, July 21, 1969. Aldrin is looking down at the systemized list of mission procedures sewn onto the surface of his left sleeve.

Why is this spacesuit soft?

On July 21, 1969, only 21 layers of fabric, most gossamer-thin, stood between the skin of Neil Armstrong and Buzz Aldrin and the lethal desolation of a lunar vacuum.

Most if not all of the functions of these layers could easily have been reduced to one or two hard layers of fiberglass or aluminum; indeed, many such "hard" suits were proposed and crafted during the era of Apollo. Both visually and organizationally simpler than their soft counterpart, such suits were arguably more efficient in their ease of movement, protectiveness, weight, and other measures.[1]

Yet these were not used. Why?

The story of the Apollo spacesuit is the surprising tale of an unexpected victory: that of Playtex, maker of bras and girdles, over the large military-industrial contractors better positioned to secure the spacesuit contract. This book tells the story of this victory, and analyzes both the Playtex suit—a 21-layer, complex assemblage—and its "hard" competitors. It is the clean lines of the latter that have traditionally captured designers' imaginations: one noted critic described the AX-3 "hard" suit as "the most beautiful designed object I have ever seen."[2]

In contrast to these "hard" suits, Playtex's A7L has traditionally been seen as a messy, almost embarrassing compromise.[3] Yet, this book argues, we are in dire need of a language to describe, and create, such elegant, adapted "softness." At a time when contemporary design discourse is turning again to a systems vernacular (albeit in a biological guise),[4] the story of the Apollo space-suit is an essential counterexample.

Against this background, the consideration of alternatives to the A7L leads us not just to the hard, military-industrial prototypes that presented alternative strategies for spacesuit design, but to a deeper kind of system proposal, that for modifying the human body itself to allow space exploration. This earlier proposal, which coined the new word "cyborg,"[5] was rooted not in sci-ence fiction but rather in the same language of systems and control—cybernetics—that was to characterize the space race as a whole.

The resulting systems of military-industrial control, first developed to build the complex nuclear missiles on which NASA's rockets depended, proved enormously successful throughout the space race. Coordinating 300,000 individuals, 20,000 contractors, and innumerable physical systems, the Apollo program's "space-age management"[6] allowed a logistical network of unprecedented complexity to reach the singular goal of landing Americans on the moon.

Yet, despite the optimism of the 1958 "cyborg" proposal, such techniques proved enduringly inadequate when it came to designing for the human body. Again and again, alternatives to the Playtex A7L would propose "engineering man for space," and, again and again, the human body aggressively resisted such encroachments. In the final reckoning, it was only when traditional engineering firms proved consistently incapable of integrating the human body into system require-ments that Playtex's proposal, and the company's intimate expertise, was accepted by NASA.

And so, in an unexpected adaptation that recalls recent work on the dynamics of evolutionary change (see Layer 20), the spacesuit worn on the moon was not developed from military-industrial expertise, but rather from underwear. And, instead of being engineered *de novo*, each Apollo suit was custom-fitted and sewn to fit its occupant alone.

1.2

The organizational and physical scale of Apollo:
personnel atop the 402-foot mobile service structure
look back at the Apollo 11 spacecraft as the tower
is moved away during a countdown demonstration
test, Kennedy Space Center, July 11, 1969.

21 LAYERS

The 21-layer A7L has a multilayered history. It was both influenced by, and influenced, fields from haute couture to medicine. To write a single, linear narrative of its creation would not only be difficult, it would deny the very quality of the object that makes it worthy of research: its open-ended complexity. The evolution of the Apollo suit provides a connective tissue to astonishingly disparate yet individually essential strands of material and visual culture, technology, and design.

This book abandons a linear tale for a collection of 21 layered essays, each addressing a topic relevant to the history of the suit, the body, and technology in the twentieth century. Holding a mirror to the adaptive, additive history of the suit itself, these layers form a sequence of 21 cross sections, or historical samples, that correspond both literally and figuratively to the 21 layers of the Apollo spacesuit.

Definition of Space

The first subsequent layer involves the discovery—both intellectually and physically—of "space" as an environment hostile to man. Occurring in the late eighteenth century, discoveries of mankind's limitations were tied together with an interest in man's own inner workings, projected to be as machine-like as the clockwork automata that provided a parallel fascination to early aviation. Space became defined, crucially, as the environment we need technology to enter.

The New Look

From this backward glance, the next layer serves as a "new look" at the postwar twentieth century. Christian Dior's fashion innovations of 1947 are used to explore a variety of themes that will be referenced throughout the book: the lessons learned by postwar business from the war effort, the focus on surface culture and the fashion cycle as engineered by a postwar consumer economy, and the onsets of new materials and techniques in the manufacture of daily life.

The New Look in Defense Planning

From the world of couture, the phrase "new look" became shorthand for a range of postwar transformations, not least the changes wrought by President Dwight Eisenhower in the management of postwar defense. Abandoning the standing army, Eisenhower committed the nation to a "New Look in Defense Planning,"[7] which emphasized nuclear technology and close collaboration between the military, industry, and academia. We are all too familiar with the legacy of this "military-industrial complex," but attention here is given to its particular institutional origins in the Atlas missile program, and the establishment therein of the standards of systems engineering and management that would serve as the institutional infrastructure of the entire Cold War space race.

Flight and Suits

From the human origins of space-age management, we turn to the humble origins of human protection against altitude, and the birth of the latex pressure suit in the barnstorming aviation culture of the interwar years. Although associated from their origin with futuristic visions, these first high-altitude "space suits" were less technocratic achievements than feats of adaptation. In 1934 such a hand-formed "tire shaped like a man" brought aviator Wiley Post to the limits of the stratosphere.

Cyborg

While primitive suits inspired by Post's were investigated by the armed forces after World War II, more synthetic notions of man in space gained credence. Notable was the neologism "cyborg," a man-machine hybrid proposed by Nathan Kline and Manfred Clynes. Clynes, a mathematician and analog computer expert, and Kline, a psychopharmacologist, proposed that instead of carrying an earthlike environment with him, the biochemistry and homeostatic functioning of man should itself be adapted for space travel. Grounded in the same theories of cybernetic control that shaped parallel advances in systems management, the cyborg concept for space exploration was studied intensively before its eventual abandonment in 1966.

Flight Suit to Space Suit

Soon after, however, came the launch of Sputnik. With the Soviet achievement, the United States accelerated its space efforts, borrowing missiles and management systems from the armed services and hastily adapting high-altitude flight suits to serve in space. In pressure suit design, military green was (literally) sprayed silver, as the iconography of science fiction shaped space flight's swiftly constructed facade.

Man in Space

Yuri Gagarin's 1961 spaceflight catalyzed that battlefield of the Cold War known as the space race, the establishing of missile superiority through the launch of man. However, the Soviet bureaucracy that had enabled the feat suppressed vital facts of the flight, including the fact that Gagarin landed only in his spacesuit, parachuting from the unstable Vostok capsule high above the earth. The particular history of spacesuits in the Soviet context is an essential and incongruous contrast to the American epic.

Bras and the Battlefield

Anticipating the space age, the International Latex Corporation (ILC), known by its consumer brand "Playtex," conducted basic research on adapting its latex expertise to pressurized suits. Initially ignored, its research gained it center stage in Apollo suit manufacture after startling advances in mobility and comfort.

JFK

In the moment of national commitment to Apollo, the artist Richard Hamilton depicted John F. Kennedy in a spacesuit, essentializing his central role in the space race, as well as his mastery of fashion and image. The image reveals an essential truth: Kennedy's pivotal decision to fund NASA's lunar effort, and to give it the urgency that shaped Apollo's realities, can be ascribed as much to the private frailties of his own body as to the oft-photographed facade it inspired.

Contractual Physiology

Instead of physiologically adapting man for space (as the cyborg model proposed), the ground-breaking research of the space age was into the real, and modest, limitations of the unaltered human body. After literally exhaustive tests, these limits were incorporated into NASA's design systems just as readily as booster thrust or orbital apogees.

Simulation

While the resulting landing on the moon shaped the visual language of our century, a greater influence arguably belongs to the virtual landscapes first crafted by NASA to allow the extensive simulation and training of lunar landings on the tight Apollo schedule. The race to the moon was a massive exercise in parallel real and simulated realities, from necessary advances in computing technology to the huge twin control rooms in Houston that alternately held real and virtual missions.

The Moon Suit Plays Football

Then, in the midst of Apollo's tense timetable, ILC was fired. In the spring of 1965, many NASA engineers, as well as top administrators, favored the military-industrial expertise of the influential Hamilton Standard/United Aircraft Conglomerate, with which ILC had initially been partnered. However, a resulting three-week competition proved the Playtex suit not only better than its military-industrial competitors, but the only suit that could reliably preserve, and extend, human abilities on the moon.

Handmade

ILC's subsequent effort was one of adaptation, both physical and organizational. Handmade assemblages of fabric, latex, and nylon were hand-sewn to minute tolerances, and custom-fitted to each astronaut. Indeed, the production of the suits met both physical and institutional barriers within the military-industrial complex, from X-ray scans to uncover errant pins, to a spirited debate on whether and how clothing sizes, as opposed to serial numbers, could be used to describe variations in individual suits.

Hard Suit 1

Subsequently, the most serious competitor to the ILC suit for later lunar missions was a hard, one-piece suit manufactured by the stratospherically successful (if ultimately disgraced) corporate conglomerate Litton Industries. Even as they failed to meet the standards for lunar use, the streamlined suits were staged by NASA as the future of space travel through the 1970s.

"We've Got a Signal!"

As the lunar landing was a media event, so was it a feat of broadcast technology—particularly at the most-viewed American news source, CBS. Designed at the same time as Kubrick's *2001* (and by the same set designer, Douglas Turnbull), the CBS lunar broadcast soundstage incorporated more simulations, asynchronous sequences, and other trappings of modern broadcasts than any other television event of its time. As well as a cultural touchstone, it was the foundation for our contemporary, 24-hour news cycle.

Hard Suit 2

Sharing the sleek geometries of CBS's soundstage, the alternative suits most beloved of design historians are the AX series of experimental suits developed by NASA's Ames Research Center. The suits provide important clues to the culture, and contours, that form the most seductive face of the space age.

Control Space

The visual seduction of space systems leads to another vision of control, provided by the Mission Control Room itself. With its implications and mythologies, the multiscreen environments of Houston offer important clues into the architecture of our own mediated milieu.

Cities and Cyborgs

These lessons regarding the body in space could be extended to larger architectural contexts through a system of analogy alone. Yet important historical links exist as well. Systems engineers and policymakers of the late 1960s sought to literally apply the lessons of Apollo to the pressing problems of cities. In these efforts they followed not the soft surface of spacesuits, but rather what they understood as the hard truths of systems engineering. And so, as with the cyborg, the subsequent failure of these efforts was as systematic as it was superficial.

21 Layers

A vital part of the 21-layered story of the A7L is provided by the idea of layering itself, and the related strategies of redundancy and interdependence. These qualities, shared by the chemical and physical reality of the A7L's 21 layers, also turn out to be essential concepts in examining the vital distinctions between natural and manmade complexity, and the qualities of robustness and fragility that define and separate them.

This final layer considers the lessons of the Playtex suit for our own age. Like the lessons of evolution that govern physical design in natural systems, they are as much in the possibilities opened up by change, as in change itself. These lessons teach us to see the future in context, as a set of possibilities and not a scripted scenario. Instead of creating spaces in our imagined, often masculine image, we may better create new looks, like Dior, out of second glances at the material we have already to hand. These adjacent possibilities offer an essential catalyst to our complex, and common, future.

1.3
Astronaut Alan Shepard shown in his
B. F. Goodrich spacesuit and Mercury
spacecraft "just prior to its being sealed,"
May 5, 1961.

ENCAPSULATED MAN

In his late-life lambasting of "technological organization," *The Pentagon of Power*, urbanist Lewis Mumford considers Alan Shepard in his Mercury spacecraft, dubbing him "Encapsulated Man." "Here," Mumford announces, "is the archetypal proto-model of Post-Historic Man, whose existence from birth to death would be conditioned by the megamachine, and made to conform, as in a space capsule, to the minimal functional requirements by an equally minimal environment—all under remote control."[8] "To survive on the moon," Mumford expands, "he must be encased in an even more heavily insulated garment."

"The astronaut's space suit," Mumford continues, "will be, figuratively speaking, the only garment that machine-processed and machine-conditioned man will wear in comfort."

As we will discover, the silvery, machine aesthetic of the Mercury spacesuit was only a surface, sprayed onto, and hiding, a much more natural interior (in the natural latex pressure bladder, literally so). With due respect to Mumford, therefore, this book makes a contrary assertion. This alternative argument can be understood through a quite different image—that of astronauts Alan Bean and Charles Conrad engaged in one of the innumerable training exercises for their Apollo 12 mission in October of 1969.

The photograph recalls the visual contrasts of *2001*'s final sequence—and, indeed, the 1968 film relied for its set designs on NASA employees recruited after post-Apollo cutbacks.[9] In a white, tightly managed interior, the two astronauts practice tracking and recording lunar samples, incongruously simulated by a field of rocks on the raised office flooring. Deep inside the Kennedy Space Center, we are presented with an organizational cross section almost as extreme as that shown earlier, on the surface of the moon. As on the moon, we see the complex, human body protected against a hostile environment. Whereas in one case it is a cold vacuum that threatens the body, in the other it is the systems and techniques of the military-industrial complex on earth. Yet in each case, the intermediating, 21-layer A7L presents a reality, not quite animate or inanimate, neither entirely natural nor manmade, that mediates the two, allowing the body to suit itself to a new historic space.

1.4

Astronaut Charles Conrad, Jr. (facing camera)
simulates picking up lunar samples, while astronaut
Alan L. Bean simulates their photographic
documentation, five weeks before their launch in
Apollo 12, October 6, 1969.

Notes

1. See for example Gary L. Harris, *The Origins and Technology of the Advanced Extravehicular Space Suit*, AAS history series, 24 (San Diego: Published for the American Astronautical Society by Univelt, 2001).

2. Michael Sorkin, "Minimums," *Village Voice*, October 13, 1987, 100.

3. See for instance the discussion of the Apollo suit program in Kenneth S. Thomas and Harold J. McMann, *US Spacesuits* (Berlin: Springer, 2006).

4. For example, Michael Hensel, Achim Menges, and Michael Weinstock of the Emergence Design Group at the Architectural Association in London lay forth an explicit agenda of "morphogenetic" architecture, relying on "data, genes, and speciation." Hensel, Menges, and Weinstock, "Emergence: Morphogenetic Design Strategies," *Architectural Design: A.D.* 74, no. 3 (May/June 2004), and "Techniques and Technologies in Morphogenetic Design," *Architectural Design: A.D.* 76, no. 2 (March/April 2006). And in a 2004 essay, Karl Chu, head of the Institute for Genetic Architecture at Columbia, declares: "We are now in a position to articulate a more comprehensive theory of architecture, one that is adequate to the demands imposed by the convergence of computation and biogenetics … a monadology of genetic architecture." Karl Chu, "Metaphysics of Genetic Architecture and Computation," *Architectural Design: A.D.* 78, no. 4 (August 2006).

5. Nathan S. Kline and Manfred Clynes, "Drugs, Space and Cybernetics: Evolution to Cyborgs," in Bernard E. Flaherty, ed., *Symposium on Psychophysiological Aspects of Space Flight* (New York: Columbia University Press, 1961).

6. A term coined by James Webb, NASA's Administrator under President John F. Kennedy. See James E. Webb, *Space Age Management: The Large-Scale Approach*, McKinsey Foundation lecture series (New York: McGraw-Hill, 1969).

7. See Saki Dockrill, *Eisenhower's New-Look National Security Policy, 1953–61* (New York: St. Martin's, 1996).

8. "To survive on the moon," Mumford expands, "he must be encased in an even more heavily insulated garment." Lewis Mumford, *The Pentagon of Power*, vol. 2 of *The Myth of the Machine* (New York: Harcourt Brace Jovanovich, 1970), 14–15.

9. Nicholas de Monchaux, interview with Frederick Ordway III, visual effects consultant to Stanley Kubrick and former assistant to Wernher von Braun, June 20, 2006 (tape recording).

Layer **2** SPACE/SUIT

2.1

The Montgolfier ascension against the facade
of the Château de la Muette.

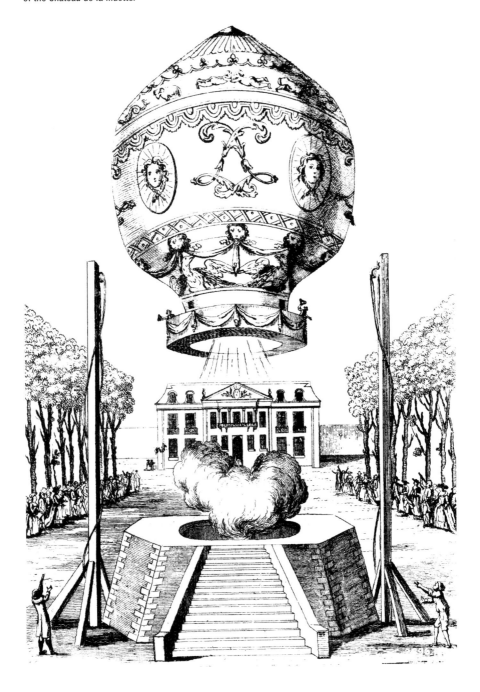

The word *spacesuit* was first coined in 1962,[1] at the dawn of America's space age. It is a compound of *space* and *suit*.

"Space," in this context, is the extra-atmospheric environment hostile to man, and for which technology is a necessary mediation. As we will see, the first glimpses of such an environment would follow quickly from the very first manned balloon flight in 1783.

Following "space," "suit" comes from the Anglo-French *siwte*, to follow (thus "pursue," as well as "sue" and "lawsuit"). Today, "suit" means a set of things—often clothing—that follow each other in style and purpose (a "suit" of cards, a "suit" of armor, and, most relevant here, a "suit" of clothes, themselves appropriate to a particular occasion, as in bathing or business).

Crucial, however, is a more subtle meaning of "suit." Just as one wears a particular suit for a particular environment—bathing, business, or space—one is also, by the action of the apparel, better "suited" to that environment. Contained here is a notion of adaptation, and control, which, at its extremes, extends beyond mere clothing. Within the history that follows—of suiting man for, and also to, space—this sense of "suit" will move us beyond surface adaptation into a mechanical mastery of the body itself. This utopian vision is an idea far older than manned flight but—as we shall see— inextricably intertwined with it.

PAST AS PROLOGUE

In 1783, the Comte Antoine de Rivarol wrote the President of the Royal Academy of Sciences on the subject of two recent popular inventions.

The first was the presentation by Étienne and Julien Montgolfier of the "globe airostatique [sic], the balloon with which mankind made his first flight."[2] This Rivarol dismissed as a fad.

Second (and more fascinating to the Comte) was a matched set of clockwork automata assembled by the provincial Abbé Mical and on their way to Paris for exhibition. The latest in a series of ever more complex humanoid clockworks presented in Paris, the heads were notable for their ability to speak using a silk larynx and leather lungs, palates, and tongues. With some difficulty, they pronounced a four-line dialog which began "Le Roi a donné la paix à l'Europe!" (concluding "O Roi adorable, père de vos peuples, leur bonheur fait voir à l'Europe la gloire de votre trône").[3] The Comte was particularly taken with the possibility that, through the use of such speaking automata, the proper pronunciation of French could be permanently fixed.

PARIS AND THE ANDROIDS

The first such *androids*, or anthropomorphic clockworks, were presented to the French court and Academy of Sciences in 1738 (the word was invented from the Greek ἀνδρο, man, + εἴδης, like). In the winter of that year, the inventor Jacques Vaucanson arrived in Paris with two androids, and a third device in the shape of a duck. Unlike previous clockwork amusements (but like the later heads of Abbé Mical), Vaucanson's machines were presented not just as public spectacle but also as serious scientific experiments.

While the androids reproduced human activities related to entertainment (playing fife and drum, as well as flute with working lips and breath), "Vaucanson's Duck" attempted a more base activity—to shit.[4] While its complex mechanism featured many movements—it could flap its wings and splash water—its chief attraction to the thousands of Parisians who paid three livres each to see it was its ability to eat kernels of grain and, after an appropriate pause, issue a smelly paste from its behind.[5] For this achievement, Voltaire hailed Vaucanson as "the new Prometheus."[6]

By the end of the eighteenth century, the techniques of Vaucanson's automata were applied, through the efforts of the automata-building Jaquet-Droz family, directly to the making of prosthetics for real human bodies. As for Vaucanson, his 1738 machines were so successful that, after a single season of their exhibition, the young inventor had acquired enough capital to leave the field. Before departing, Vaucanson drafted an elaborate proposal to Louis XV for the simulation of the human circulatory system, using latex from Guyana. This proposal was not advanced, however, and the inventor instead devoted his later career to fabric manufacturing, most notably the automatic pattern-weaving process that now bears the name of Joseph-Marie Jacquard (whose 1804–1805 model improved Vaucanson's 1747 design).[7]

ATMOSPHERES AND FABRIC

It was underneath an envelope of hand-woven silk fabric that the Montgolfier brothers successfully raised the first human "aeronauts" on November 21, 1783.[8] Although there was some confusion over the nature of the "air" that sustained the flight of the Montgolfiers' balloon, the social and scientific atmosphere that drove it was the same as that which produced Vaucanson's androids.

In a delicate balance between state sponsorship, private profit through subscription to events, and the potential for academic fame given through the imprimatur of the Academy of Sciences, eighteenth-century France supported an extensive scientific *demi-monde*.[9] In this context, popular scientists and purveyors of demonstrations existed on a spectrum from those like Vaucanson (and later the Montgolfiers) whose public demonstrations were of a caliber to interest and engage the Academy and royal patronage, to those whose efforts simply allowed a living to be made through showy demonstrations and public events (among this latter number was the doomed revolutionary Paul Marat). The resulting *physique expérimentale* included exercises in static electricity, magnetism, and hydraulics as well as automata and balloons, and placed equal emphasis on scale, spectacle, and scientific progress. The enormous balloon ascensions of 1780s Paris were symptoms of such "popular" science; quintessentially spectacular, they were not immediately thought to have any practical use at all.

The ascensions were celebrated at every level of culture, "absorbing French society as nothing else in the decade before the revolution."[10] The Marquis d'Arlandes and Pilâtre de Rozier, two aristocrats who had engineered their substitution for the planned cargo of convicts on the maiden voyage of the Montgolfier balloon, were admitted to the Academy of Sciences by acclamation (the Montgolfiers themselves ascended through a traditional ballot). At the same time, the physical figure of the balloon became a visual icon in culture high and low. The architect A. T. L. Vaudoyer published the design of a "house for a cosmopolite" in the shape of a balloon in 1785, in parallel to balloon-inspired designs by Boullée (the cenotaph for Newton) in 1784 and a series of architectural essays on the balloon by Claude-Nicolas Ledoux. Beyond the high arts, balloon motifs graced liquors ("crème aérostatique"), hairdos, wallpaper, bed warmers, barometers, and chamber pots.

Not least of all, the motif dominated fashion, both male and female.[11] The shape referenced in these designs, however, was not the November 19 Montgolfier balloon (which resembled an obelisk in elevation), but the seductively perfect sphere of a much smaller and more popularly successful hydrogen balloon launched on December 1, 1783. It is an ascension worthy of particular regard.

At the invitation of the Madame de Polignac, governess of the royal children, the Montgolfiers' November 19 flight was held on the outskirts of Paris. (Benjamin Franklin was in attendance, and described the crowd as a "vast concourse of Gentry.")[12] The December 1 hydrogen balloon launch, by comparison, attracted some 400,000 Parisians to the area around the Tuileries—more than half the population of the city.[13]

The architect of this second manned ascension, Jacques Alexandre César Charles, was one of the most noted scientific showmen of his time, with a *cabinet de physique*—the physical tools for demonstrations and teaching—determined to be "the most complete and curious that Europe possessed."[14]

While Charles received permission from the royal court to use the confined space of the Tuileries for his launch, the ascension itself was financed through a system of public subscription. An early donation of four louis (96 livres) entitled the subscriber and 30 guests to enter an enclosure in which the flight was prepared. Individual tickets were sold for six livres each (more than a week's wages for an average worker), while a three-livre billet allowed the bearer to enter the Tuileries only. On Friday, November 27, 1783, when receipts for these tickets could be exchanged for the actual article, the crush of crowds was so great that the launch was postponed to the following Monday.

Despite its popular origins (Étienne Montgolfier dismissed subscription-based science, observing that it left one "accountable to the [public]"),[15] the Charles launch was a technological *tour de force*. The craft, with a basket suspended evenly from a gas envelope and an elaborate system of ballast and gas release, introduced a system of aerial navigation that would survive into the twenty-first century. The less tangible technical achievements of the Charles balloon were a series of insights into the curious notion of "air," which sustained balloons and, it became clear, their pilots as well.

AIR AND SPACE

The French and English "air" makes its way from the Greek ἀήρ—to blow, or breathe. By definition, air is that which we inhale and exhale (thus the operatic "aria" as well as dreaded "malaria"). At the time of the first balloon ascensions, the word "air" was used to describe a substance that a modern audience would recognize as a "gas"—elemental and atmospheric. But it was only in the late eighteenth century that air had been shown to be made of constituent gases, which became known as different types of "air." The seventeenth-century chemist Robert Boyle coined the term "facetious air"[16] to describe those gases not ordinarily found in the atmosphere. By the late eighteenth century, many such airs were known, including marine air (muratic acid gas), fixed air (carbonic acid gas), and inflammable air (hydrogen). Those substances found to be nonfacetious—naturally occurring in the atmosphere—consisted chiefly of phlogisticated air (nitrogen gas) and, in a smaller but crucial proportion, vital air, or oxygen.

When the Montgolfier brothers sent up their first trial balloons in the summer and fall of 1783,[17] they burned not just straw but also "Old Shoes, Rabbits Skins, and other animal substances."[18] Despite advice from the Academy of Sciences to the contrary, Joseph Montgolfier maintained a belief that no fewer than four "principal agents" contributed to the balloon's flight. Along with the loss of density in the captive air due to heating, Montgolfier cited "vapeurs aqueuses et huileuses" from the straw and animal products burned, the "inflammable air" (hydrogen) produced by these same substances, and the "electrical fluid" he insisted charged the balloon's interior.[19]

Professor Charles, by contrast, engaged in a more scientific approach to gases. For an un-manned test on August 27, and leading up to his ascension on December 1, 1783, the "experimen-tal philosopher" designed an apparatus to produce large quantities of hydrogen. Using numerous full and half barrels, leather tubes, iron filings, and "oil of vitriol" (sulfuric acid), the preparation of gas preceded the balloon launch by several days. Inflated by the expanding gas, the sphere was constructed from alternating swathes of red and yellow silk.[20]

Even with a functioning craft, doubt raged before and after the first ascensions as to whether high altitudes, reached by balloon, could sustain life.[21] As preparations were made for the Montgolfier ascension, several letters to the brothers warned of the potential effects of "air auréale."[22] The Montgolfiers were advised to enclose the balloonist in a tight wooden box "resembling a coffin," or alternatively create an atmosphere of breathable air by burning substances "fecund in the oil and earthly principles," such as almonds or dried mushrooms.[23]

After landing 15 kilometers from Paris with his copilot in the late afternoon of December 1, 1783, Professor Charles, curious about higher altitudes, commanded that the balloon be sent up again with him alone as passenger. Thus lightened, the balloon shot to 6,000 feet within ten minutes. The sunset, already ending, became visible once again ("I had the pleasure of seeing the sun set twice in the same day," Charles rejoiced on landing).[24] So distracted, Charles allowed the balloon to ascend to more than 10,000 feet in altitude until, brought alert by "intense fright and unease, cold, and earaches,"[25] he quickly released hydrogen to descend. Chastened, Charles did not fly again.

ASCENDING ALLIGATORS

After Charles's ascension, the first "true" scientific attempt at high-altitude flight was made on September 5, 1862, when James Glaisher of the Greenwich Observatory accompanied Henry T. Coxwell, a balloonist with decades of experience, on an ascent from Wolverhampton. They left the ground in fog at 1:03 p.m., carrying small "container balloons" of oxygen, from which they hoped to gain sustenance at high altitude. By 1:53 p.m. they had reached 29,000 feet, and difficulties arose. The balloon held a cage of pigeons as test subjects, and several had been dropped at 15,700, 21,000, and 26,000 feet—"falling like leaves." By 29,000 feet, the two remaining pigeons had died. By the time the balloon had reached 33,000 feet, Glaisher had become extremely drowsy, felt his arms grow numb, and saw his vision fail. In an attempt to revive himself, he reached for a bottle of wine, only to discover that he was paralyzed. He tried to call to his companion, who was climbing the basket to try to release gas from the balloon, and found he could not make a sound. Coxwell, in the rigging, felt his movement slipping away, and was finally only able to pull the line to the valve with his teeth. The balloon descended quickly, and the two men were saved.

After Glaisher and Coxwell's brush with death, a subsequent expedition financed by the French Academy relied on a primitive altitude chamber constructed by physician Paul Bert to train its crew. Joseph Croce-Spinelli, Théodore Sivel, and Gaston Tissandier practiced the use of pressurized oxygen canisters, periodically self-administered, over several months of chamber trials. Their subsequent flight rose from Paris on the morning of April 15, 1875. Launched at 11:35 a.m., the balloon reached a height of 16,000 feet soon after 1:00 in the afternoon. At approximately 1:30, with the balloon still rising, Tissandier attempted to announce to his colleagues a barometer reading of 26,000 feet. But he could not. Suffering the effects of oxygen deprivation, he was deprived of speech, and soon lost consciousness altogether.

Except for a brief glimpse of lucidity—in which he recalled observing Croce heave an enormous metal "aspirator" out of the plummeting basket—Tissandier did not awake until 3:15 p.m. Again, the balloon was descending rapidly. Attempting to rouse his companions, he saw their eyes glazed and mouths filled with "bloody froth." They were dead.

In the remaining minutes of his ill-fated expedition, Tissandier threw all but his companions' bodies overboard, and made a crash landing near the village of Ciron shortly after 4:00 p.m. Believing himself about to die like his companions, he collapsed. By the next day, however, he had recovered sufficiently to deliver a full report to the Academy. A subsequent analysis of the flight would conclude that relying on an aviator to consciously administer oxygen to himself was inherently flawed. The strategy ignored the tendency of oxygen deprivation to reduce the judgment and mobility necessary for its own execution.

2.2
A popular illustration featuring Glaisher and Coxwell as they succumb to the effects of altitude, September 1862.

UP INTO THE JAZZ AGE

When U.S. Army Air Corps Captain Hawthorne C. Gray rose into the Illinois sky in 1927, he was one of the first since Tissandier to seriously challenge human endurance in the upper atmosphere. Gray's equipment would seem to be an age away from that of earlier expeditions. He was equipped by technicians at the Air Corps' airship school at Scott Field to endure all known hazards of high-altitude flight. Fully suited, he wore 57 pounds of protective clothing, covering even his face and hands, and was supplied with oxygen through a heated assembly modified from a gas mask, hooked to a series of redundant tanks. What provided the greatest contrast with his predecessors, however, were the dazzling licks of St. Louis jazz floating through his balloon basket, delivered via the newly termed "airwaves."

It was the invention of modern radio itself that sparked the renewed interest in the atmosphere's limits. When, in the first years of the twentieth century, radio signals were successfully broadcast over large distances, it was theorized that the "conducting" layer of air reflected the high-frequency electrical signals back to earth. In 1924, Edward Appleton proved the existence and altitude of such an "ionosphere" using BBC radio equipment. Years later, he was awarded the Nobel Prize for the achievement, the citation lauding him as an "explorer of the unknown regions which form a border between our earth and the universe."[26]

Against this background, Hawthorne Gray set out in November 1927 for the last of three high-altitude flights. His craft would soar above 44,000 feet, but an inaccurate estimate of his breathing supply's longevity would leave Gray dead long before the balloon's landing at 5:20 p.m. "Undoubtedly," a posthumous award of the Distinguished Flying Cross concluded, "his courage was greater than his supply of oxygen."[27]

The last flight of Hawthorne Gray ends the chronicle of balloonists' naked vulnerability to the upper atmosphere opened by J. A. C. Charles's 1781 discomfort. Yet it was far from the end of adventures in altitude. Fueled in part by the very "broadcast media"[28] whose transmission first revealed the ionosphere, the race for height continued.

Gray's demise initially slowed U.S. efforts, and Frenchman Auguste Piccard (sponsored by the same Academy of Sciences that commissioned Charles's and Tissandier's flights) set a new altitude mark in a pressurized gondola in 1931. With an invitation to ascend at Chicago's Century of Progress exhibition in 1932, Piccard's work was incorporated into renewed U.S. efforts. Competition in this new era came from the upstart Soviet Union, which flew its first record-setting flight in 1933. Like the Piccard balloon, however, the Soviet device did not leave its aeronauts exposed to the elements, and enclosed them instead in a rigid metal sphere. The record-setting flights of the 1930s would reveal much of the upper atmosphere, but their pilots, like nineteenth-century submariners, would observe the new realm only through miniature portholes and scientific instruments, lodged in spherical prisons whose perfect shapes echoed the balloon-inspired architecture of Claude-Nicolas Ledoux. The apex of such flights was the National Geographic Society and Army Air Forces' *Explorer II*, launched from a dedicated "Stratobowl" in South Dakota, and the subject of hundreds of news articles. (Particularly fêted by the press were the World War I Army Air Corps heroes substituted for the Frenchman Piccard as passengers.)

Explorer II's destination was the "stratosphere," a sphere of air not plagued by ground-level currents, but instead divided into steady layers (strata), which, as distinguished from ground-

2.3

Hawthorne Gray (at right, in leather and
sheepskin flight clothing) prepares for a test
ascension in the balloon gondola at left.

level air, increase in temperature with altitude. This is due to the realm's relative proximity to the unshielded sun, the effects of which help produce a series of increasingly alien environments further above. As they were revealed by balloon flights throughout the 1930s, the strange details of such new aerial "spheres" were reported to the public with delight. "We live in the mere dregs of the atmosphere," proposed the *New York Times*, "turbulent and murky in comparison with that serenity [aloft]."[29] In succession, the public learned not just of the stratosphere, but the succeeding mesosphere, thermosphere, and exosphere, defined by different relationships between altitude and temperature. Sprinkled among these as new discoveries were not one but two separate ionospheres (at 60 and 150 miles of altitude), as well as the thin and precious ozone layer, whose ionized oxygen blocks damaging radiation from the earth's surface.[30]

SPACE

As our notions of "air" became increasingly rarefied (the last zone of the atmosphere, the sparse "exosphere," would turn out to extend many thousands of miles above the earth's surface), a new word entered the popular lexicon—"space." This use of the word came to mean not simply our everyday experience of dimension, but—especially in its capitalized form—an outer "Space" that lies in explicit contrast to the everyday. In a literary context, this capitalized "Space" appears as far back as John Milton. No lesser mouth than Satan's drops the phrase as he describes the void from which planets (and later, man) arise:

> "Space may produce new Worlds; whereof so rife
> There went a fame in Heav'n that [God] ere long
> Intended to create, and therein plant
> A generation, whom his choice regard
> Should favor equal to the Sons of Heaven."[31]

When later Satan travels from hell to a newly created realm and gazes at our distant, "pendant world" of Earth, he is supported by the "the emptier waste, resembling Air."

As a practical matter, the definition of this unearthly realm depended on one's perspective.

For astronomers, bringing their telescopes to higher and higher altitudes in the early twentieth century, "space" would begin at approximately 5,000,000 feet of altitude, where the atmosphere would offer no distortion to a star's thin stream of light.

For aviators, "space" came to begin at roughly 250,000 feet above earth, where air no longer supports winged flight. (This corresponds with a definition of space held by electrical engineers; at 250,000 feet, the atmosphere is no different from the interior of an earthly vacuum tube.)

For early rocket engineers such as Robert Goddard (who fired his first liquid-fueled device in 1926), "space" came to mean an altitude of approximately 100,000 feet, where the atmosphere no longer restricts the expansion of gases from a rocket's engine, allowing optimum thrust.

For the sake of aviation speed and altitude records, after the Second World War the Hungarian scientist Theodore von Kármán proposed an arbitrary line separating "Aviation" from "Astronautics," corresponding to the relative influence of atmospheric and orbital effects. (The altitude approximated 100 kilometers, and was adopted as such.)[32]

For doctors, and the human bodies they examine, "space" is the most restrictive of all. Above three miles, or 15,000 feet of altitude, unacclimatized human beings need oxygen to survive longer than several hours, and above seven miles, or 35,000 feet, a pressurized environment is necessary to avoid eventual unconsciousness and death.

SPACE AND MAN

Contrasting this fragility, the clockwork androids of the eighteenth century offered a tantalizing possibility to scientists and philosophers: that man, thinking, feeling, and possessing a soul, could be modeled, and modified, as a machine. Citing Vaucanson's automata, Julien Offray de La Mettrie, an Enlightenment doctor and philosopher, wrote in 1748: "Vaucanson, who needed more skill for making his flute player than for making his duck, would have needed still more to make a talking man, a mechanism no longer to be regarded as impossible." Underlining the interdependence of thought, feeling, and physical reality, La Mettrie claims for the whole self, body and soul, an innate identity as a machine, "a collection of springs, which wind each other up."[33] Implicit in his claims, offered against the growing technological mastery of Enlightenment science, there appears the imminent prospect of mastery over the body, and thus over life itself. Here was the dawn of a new mankind.

Surviving only a few years into the nineteenth century, however, the clockwork android quickly lost its scientific stature and sophistication (the technology bastardized to make figures of entertainment). As new complications of anatomy and new theories of physiology took hold, the idea of man as literal master of his own machinery slipped, for a time, out of view. The balloon, by contrast, soared skyward. The balloon's enslavement to the winds moderated the transformative changes in infrastructure and distance that were initially predicted by contemporary observers. But the potential of the craft to fuel the public imagination with constant increases in height, spectacle, and scale would ensure the long-term viability of early experimental craft.

Which brings an irony. From its outset, manned aviation took the body to its fragile limits, exposing pioneers to discomfort, disaster, and death. Yet at the same time, the image of the airborne man came to convey not only a mastery of the earthly realm, but a marvelous separation from it and removal to a new kind of humanity. When the *New York Times* fêted high-altitude balloonists in the 1930s for slipping beyond our "murky realm," it was in the same spirit as the American Charles Ferson Durant, dressed in top hat and tails, dropped copies of his "First American Aeronauts Address" above Manhattan on September 9, 1830: "Good bye to you—people of Earth / I am soaring to regions above you."[34]

Aided by the later efforts of aeronauts, "space" grew to describe a realm defined by its increasingly inaccessibility. But by ascending close to it through flight, being *suited* to it by his craft and garb, the adventurer himself became imbued with this most unearthly of qualities. By ascending above the normal limits of physiology he became, literally, "super-" human. At the insistence of Louis XVI, the Montgolfiers assumed a balloon-decked crest with the motto *Sic itur ad astra*, "Thus we reach the stars."

On Bastille Day, 1789, a revolutionary mob burst into J. A. C. Charles's honorary apartment at the Louvre, to ransack it as they had the rest of the palace. Yet, on recognizing the 1783 balloon hanging from the rafters, they judged his achievements to be so beyond the concerns of the day that they not only spared him and his apartment, but petitioned him for a seminar that lasted far into the night.[35]

The popular character of the superman aeronaut—still as powerful when the pilots of *Explorer II* were entertained by Franklin Roosevelt[36] as when Charles escaped the political activists of another era—recalls the mastery of mankind's sphere celebrated by eighteenth-century technologists. Yet, as made clear by Charles's own aerial discomfort, the man "above earth" was very much still a fragile being, by definition unsuited for his realm. After the disastrous flights of Hawthorne Gray, the preparations made to occupy high-altitude environments no longer aspired to anthropomorphic "suiting" at all—instead focusing on rigid containers. Recalling the protective "coffin" recommended to the Montgolfiers, such craft brought all the dearth of sensation and engagement that the word implies. The cabin-bound flier may have technically entered another realm, but he was dead to it.

SPACESUIT

It was to the notion of *living* in space, then—operating in space as one might on the earth's surface—that the word "spacesuit" came to apply. More than just *suit* as ensemble, the spacesuit *suited* its wearer to his lofty sphere. And in so doing it brought together two ideas—man as master of his human machinery, and man as hero of a heavenly realm—that we have heretofore regarded separately.

The literal origins of the spacesuit were not based in such concerns, but rather in the practical needs of a new generation of aerial craft-powered airplanes. Yet almost as quickly as they were invented for practical reasons, the pressurized high-altitude suit—progenitor of the modern spacesuit—was celebrated for its unearthly quality. Within decades, as the technology of powered flights opened higher and higher realms to mankind, the spacesuit would stand as icon of a transformed, exploring man.

For all its superhuman iconography, however, the twentieth-century spacesuit would never transform human biology. This was not for lack of trying, as evidenced by the origin of words like "bionic" and "cyborg" within NASA studies. Yet in "suiting" man to an environment defined by its hostility to him, the *spacesuit* itself would come to play a central role in discussions of mankind made and remade by technology, earthly and, it sometimes seemed, divine.[37]

Notes

1. "G suits are not to be confused with pressure suits (or now spacesuits) which the Astronaut wears during space flight to maintain atmospheric pressure at high altitudes." John Glenn et al., *Into Orbit* (London: Cassell, 1962), 244. The *Oxford English Dictionary*, 2nd edn. (1989), references this mention as the first use of the compound "spacesuit." According to the same definition, "space suit" as a separate phrase dates from 1929.

2. Comte Antoine de Rivarol, *Lettre écrite à Monsieur le Président de ***. Sur le globe airostatique, sur les Têtes parlantes et sur l'état présent de l'opinion à Paris* (Paris, 1783).

3.   Intermediary lines are: "Le paix corrone le Roi de gloire. Et le paix fait le bonheur de peuples."

4.   It would be a mistake, however, to assume that the humanoid companions to the duck were less complex. Unlike previous musical androids, the flute and pipe players played their instruments with flexible and artificial lungs, lips, and fingers, the design of which was the subject of a lengthy article submitted by Vaucanson to the French Academy of Sciences.

5.   See Jessica Riskin, "The Defecating Duck, or, the Ambiguous Origins of Artificial Life," *Critical Inquiry* 39 (Summer 2003): 599, and a related article by the same author, "Eighteenth Century Wetware," *Representations* 83 (Summer 2003): 97.

6.   Voltaire, "Discours en vers sur l'homme (1738)," in *Oeuvres complètes*, vol. 9 (Paris, 1877), 420; referenced in Riskin, "The Defecating Duck." The duck was later revealed by the magician and engineer Houdini to be a fraud; its rectum was preloaded with a malodorous paste before each demonstration.

7.   Herman Blum, *The Loom Has a Brain: The Story of the Jacquard Weaver's Art* (Philadelphia: Craftex Mills, 1959), 33.

8.   It was only due to the generosity of seamstresses in the crowd outside Paris that the craft left the ground at all. After a mishap with a tethered test earlier in the morning, the voluminous fabric envelope seemed damaged beyond repair. Women in the audience volunteered their needles, however, and the balloon was able to take flight later in the afternoon. A formal account of the flight and its preparations was published in the *Journal de Paris*, November 24, 1783, 1340–1341.

9.   For a description of this atmosphere in the context of balloon flight, see James Martin Hull, "The Balloon Craze in France, 1783–1799: A Study in Popular Science," PhD diss., Vanderbilt University, 1982, available through UMI.

10.   Ibid., 150.

11.   Ibid., 151.

12.   Benjamin Franklin to J. Banks, November 21, 1783, in Abbot Lawrence Rotch, *Benjamin Franklin and the First Balloons* (Worcester, MA: Davis Press, 1907), 9.

13.   Tom D. Crouch, *The Eagle Aloft: Two Centuries of the Balloon in America* (Washington, D.C.: Smithsonian Institution Press, 1983), 34–35.

14.   Archives Nationales de France, F17 1219, ds. 12, "Rapport présenté au Ministre de l'Intérieur, 5 floréal an 7."

15.   Letter of January 19, 1784, quoted in Léon Rostaing, *La famille Montgolfier, ses alliances, ses descendants* (Lyon: Rey, 1910), 298.

16.   See Robert Boyle, *A Continuation of New Experiments, Physico-Mechanical, Touching the Spring and the Weight of the Air, and their Effects* (Oxford: H. Hall for R. Davis, 1669).

17.   The French phrase "trial balloon," *ballon d'essai*, appeared in French popular writings at this time, used in the sense of the current English phrase (which entered popular use only in the twentieth century).

18.   Letter from Peter Jay Munro to his uncle John Jay, October 16, 1783. MS. 42.315.132, Museum of the City of New York; quoted in Crouch, *The Eagle Aloft*, 31.

19.   FM VIII-23, and Joseph Montgolfier, "Discours de M. de Montgolfier."

20.   Despite Charles's innovations, it should not be deduced that the technique and behavior of his balloon's "airs" were properly understood even by later pilots. When Pilâtre de Rozier, an aeronaut of

the first Montgolfier flight, set off to cross the English Channel in a balloon from Calais in 1785, it was apparently not clear to him or his backers that the balloon design—which suspended a hot-air device underneath a gasbag of hydrogen—would result in an explosion as the hydrogen bag was heated. Yet so events transpired and, soon after launch, Rozier and his companion became the first fatalities of the balloon age.

21. A modern sensibility would instantly equate the air of altitudes reached through balloons to the mountain air reached by missionaries and explorers by this time. It is clear from discussions of the time, however, that the atmosphere reached by balloon was judged, through its emptiness and previous lack of occupation, to be qualitatively different from an Alpine air.

22. FM XXI-29, dated Paris, December 2, 1783, quoted in Hull, "The Balloon Craze in France," 114.

23. Hull, "The Balloon Craze in France," 145.

24. *Journal de Paris*, December 14, 1783, 1431, in Hull, "The Balloon Craze in France," 146.

25. Melvin B. Zifsein, *Flight: A Panorama of Aviation* (New York: Pantheon Books, 1981), 9.

26. From the speech of introduction by Arne Tiselius, Vice-President of the Royal Academy of Sciences, Nobel Prize Ceremony, December 10, 1947, in Arne Holmberg, ed., *Les Prix Nobel en 1947* (Stockholm: Nobel Foundation, 1948).

27. "Army Flying Cross Awarded for Five," *New York Times*, February 21, 1928, 13.

28. Before the 1920s, "broadcast" was a phrase most often used to describe a wide dispersal of agricultural seed, and "media" still referred to a biological barrier, as a vein or septum. Its modern usage, describing the interposition of "mass media" between one's body and world events, first appeared 1923. (*Oxford English Dictionary*, 2nd edn.)

29. "Into the Stratosphere," *New York Times*, June 9, 1935, E8.

30. See D. DeVorkin, *Race to the Stratosphere: Manned Scientific Ballooning in America* (New York: Springer-Verlag, 1989), 27–29.

31. Milton, *Paradise Lost*, Book I, line 650. Cited by the *Oxford English Dictionary* as the modern "Space." Also Book II, lines 1052 and 1045.

32. The altitude is now named the von Kármán's line in the scientist's honor. Dr. S. Sanz Fernández de Córdoba, ICARE President, "100 km. ALTITUDE BOUNDARY FOR ASTRONAUTICS," website of the Federation International Aeronautique, Astronautics Records Commission, <www.fai.org/astronautics/100km.asp> (accessed April 1, 2006).

33. Julien Offray de La Mettrie, *Man a Machine*, trans. Gertrude Bussey and M. W. Calkins (La Salle, Illinois: Open Court, 1912).

34. Charles Ferson Durant Scrapbook, box 233, AIAA History Collection, Manuscript Division, Library of Congress; referenced by Crouch, *The Eagle Aloft*, 148.

35. See Institut, ms. 2038, f. 151–152, and Hippolyte Carnot, *Mémoires sur Carnot par son fils* (Paris, 1863), vol. 1, 251.

36. "President Greets Balloon's Pilots," *New York Times*, November 15, 1935, 8.

37. In 1816 (after viewing several eighteenth-century Jaquet-Droz mannequins in Geneva), Mary Shelley drafted the first novel of man-created superman in the early nineteenth century. Echoing Voltaire's description of Vaucanson, it was entitled *Frankenstein or the Modern Prometheus*. The first 1818 edition features an epigram from *Paradise Lost*: "Did I request thee, Maker, from my clay/ To mould me man? Did I solicit thee/From darkness to promote me?"

Layer **3** THE NEW LOOK

On February 12, 1947, *Harper's Bazaar* editor Carmel Snow moved quickly through a crowded audience in Christian Dior's small Paris atelier and congratulated the designer on his first solo collection. The group of dresses, with their flared hem, wasp waist, and tight bodices, has become known as the "New Look."

The phrase was coined by Ms. Snow, who quickly exclaimed to Dior: "It's quite a revolution, dear Christian! Your dresses have such a new look!"[1] Quite removed from the reality outside the studio—that same morning, the daily ration of bread in the city of Paris had been lowered from 350 grams per person to 200—the dresses were extravagant in their use of materials. Using up to twenty yards of fabric, the clothes boldly reconfigured the feminine silhouette. Emphasizing a smaller waist and longer skirt, the new clothing eschewed the practicalities of wartime work and material deprivation. Yet simply wearing one of the New Look dresses was work in itself; the heaviest weighed sixty pounds and all required a tight-girdled waist, limiting breath and mobility.

For Dior, and then the world, the New Look was a sensation. Within ten days, the fledgling fashion house had fulfilled its financial goals for the year.[2] Within months, through the intervention of Carmel Snow and others, the New Look became a media darling, altering tastes—as well as hemlines and waistlines—across Europe and America. And by October, for socialite Nancy Mitford, it had become tiresome. She wrote to her sister Diana: "Is your coat black? Mine is really a slight bore because of all the staring, and yesterday a strange woman said would I excuse her asking but does it come from Dior?"[3]

The Fashion System

Within our current story, the New Look is notable for several reasons.

First, the New Look of Dior marked the end of a closed world of couture fashion, and the birth of what French critic and semiologist Roland Barthes later termed the "Fashion *system*."[4] As Barthes explained, such a system—a series of networked relationships between designers, magazines, and consumers—subsumed the singular objects of fashion, such as dresses, into multiple layers of meaning, acquired through mass-produced photographs and their captioned commentary.

Such a system meant that, in reality, very few women wore Dior's couture creations; instead, the world experienced the New Look through thousands of column-inches devoted to the phenomenon from 1947 onward. (In 1953 alone, Dior's publicist recorded 1,300 separate articles on Dior in the American press.)[5] Like the Apollo spacesuit at the core of this narrative, the massive impact of the New Look dresses was one of image. Their effect on culture was in inverse proportion to their intimate scale. Or, as Dior himself commented, it was a "bloodless, (but inky!) revolution."[6]

3.1
1947 New Look dress "Bar," modeled by Renee.

FASHIONING ORGANIZATION

At the level of organization, Dior's New Look marked the beginning of what was a vital supporting framework of the fashion system—the business operation that would transform Dior into a global brand. As Dior's own biography makes clear, this grafting of physical fashion onto an organizational infrastructure was not a Parisian invention, but imported from the United States.

In September 1947, following the worldwide notoriety of the New Look dresses, Stanley Marcus, founder of the Dallas-based Nieman Marcus chain, invited Dior to an awards dinner, the "Fashion Oscars." Located in Dallas, the ceremony would be followed by a month-long tour of the United States, accompanied by Marcus. While some parts of the trip reflected Dior's interests in architecture and urbanism (he reveled in the "thrusting obelisks" of New York and embraced "the slums")[7] it was substantially a seminar in the business of fashion, with visits to department stores and clothing factories across the country.

For Dior, it was a "course in business school."[8] As a direct result of his experiences in America, Dior became the first couture fashion house offering a ladder of products at different price levels, sustained through a series of outlets, international franchises and labeling deals that augmented the semiotic "system" of fashion with a wide selection of products and a global network of shopfronts. Of final interest to Dior was the way in which architectural innovations in the United States allowed continued work throughout the seasons. Upon his return, Dior installed the first Carrier air conditioner in Paris at his atelier.[9]

NEW NEW LOOKS

A third impact of the New Look goes beyond the semiotics of fashion into the language of global culture. Here, the phrase "new look" came to stand not so much for the shifting shapes of Dior's 1947 dresses, but for an architecture even more ephemeral, that of postwar reality itself. From dressing gowns to government policy, the phrase "new look" moved beyond even the colossal power of postwar marketing to describe all those jarring elements that the media-rich, newly industrialized United States found itself confronting after 1945.

In February of 1949 alone, the *New York Times* discussed a "New Look" in materials (plastic), criminal justice (new streamlined courts), infrastructure (new highways), and politics (changes in the structure of the Republican Party). "New Look" had become shorthand not just for fashion, but for all those things in the momentous shifts of postwar life that, as Dior later observed, "we believe to be promising, [but] disconcert the public at first."[10]

3.2
Dior selecting fabrics in his atelier, 1957.

Books for
Bill —
T.

INTIMATE ARCHITECTURE

The American profession of architecture, which turned to prewar European modernism after the Second World War, seems to be one of the few facets of life to escape the "New Look" sobriquet (even as it widely adopted the new "international" style). Yet it was to the language of architecture that Dior himself turned to describe his own innovations. "In order to satisfy my love of architecture and clear-cut design," he wrote in his 1957 autobiography, "I wanted to employ quite a different technique in fashioning my clothes, from the methods then in use—I wanted them to be constructed like buildings." Dior particularly emphasizes the alterations and enhancements in the female form that the New Look proposed: "I emphasized the width of the hips, and gave the bust its true prominence."[11]

"Pare down, trim, prune, refine, strip away, and adopt the New Look," *Elle* magazine encouraged its readers in October 1948. Yet, far from a minimal affair, the surface elegance of the New Look hid considerable layered supports. Dior's "enhancing" of the female form was also a careful restriction of it. Perhaps architectural but also unnatural, the New Look silhouette was made possible only through the structural work of foundations—the so-called "foundation garments" of a structured bra and fitted girdle that made the streamlining of the body possible.

Only two decades after the New Look, when an American man stood on the surface of the moon and his photograph beamed round the globe, his garments would be as restrictive as a 1940s rubber girdle—and made from the same material. In an irony that much of this book will contemplate, the spacesuit of 1969 and the most popular girdle of 1949 shared a common manufacturer—the International Latex Corporation, of the consumer brand Playtex. Each, in its own way, is a new look of the postwar era, and each presents an architecture that both distorts and adapts the body to shape the technical and mediated realities of the postwar age.

A final connection is one not merely of resonance, but of causality. As we will see throughout our multilayered narrative, the expertise of Playtex in the craftsmanship and manufacture of intimate architecture would see its couturier-like sewing workshops produce a lunar spacesuit that bested the best efforts of much larger and better-funded military-industrial proposals. Yet the exploratory research that laid the foundations for this unlikely victory was funded directly by the company's flush fortunes after the Second World War, when the media's embrace of the New Look ensured that a "foundation garment" lay underneath the new age's public silhouette.

"Without foundations," Dior himself observed, "there can be no fashion."[12]

3.3

Backstage at the Spring/Summer 1953 collection, "May" dress from the Tulipe line, modeled by Alla.

Notes

1. Marie-France Pochna, *Christian Dior: The Man Who Made the World Look New*, trans. Joanna Savill (New York: Arcade Publishing, 1996), 131–135.

2. Farid Chenoune, *Dior*, trans. Barbara Mellor (New York: Assouline, 2007), 12.

3. Nancy Mitford and Charlotte Mosley, *Love from Nancy: The Letters of Nancy Mitford* (London: Hodder & Stoughton, 1993).

4. See Roland Barthes, *The Fashion System*, trans. Matthew Ward and Richard Howard (New York: Farrar Straus & Giroux, 1983), originally published in French as *Système de la mode* (Paris: Éditions du Seuil, 1967).

5. Pochna, *Christian Dior*, 170–171.

6. Christian Dior, *Dior by Dior: The Autobiography of Christian Dior*, trans. Antonia Fraser (London: Weidenfeld & Nicolson, 1957, reprinted London: V&A Publications, 2007), viii.

7. Pochna, *Christian Dior*, 195.

8. Ibid., 207.

9. Ibid., 208.

10. Dior, *Dior by Dior*, 152.

11. Ibid., 23.

12. Richard Martin and Harold Koda, *Infra-Apparel* (New York: Metropolitan Museum of Art, 1993), 21.

Layer **4** THE NEW LOOK IN DEFENSE
PLANNING

4.1

"High Ranking Personnel" observe the Greenhouse
series of nuclear tests in the Pacific proving
grounds, Eniwetok Atoll, 1951.

Some six years after Christian Dior's debut collection, on December 14, 1953, Admiral Arthur W. Radford gave a speech to the National Press Club entitled "The 'New Look' in Defense Planning."

Admiral Radford was President Eisenhower's chairman of the Joint Chiefs of Staff, and the speech outlined a series of debates that had consumed the Eisenhower administration's defense planning throughout its first year. They culminated in the National Security Memorandum NSC 162/2 of October 30, 1953, which outlined a substantial change in American defense policy. After Radford's speech, the policy became known as "Eisenhower's New Look."[1]

It was not just novelty that fueled the allusion. Recalling the substance as well as the style of Dior's 1947 collection, Eisenhower's New Look was a substantial refashioning of the defense establishment. Instead of rearranging fabric, however, Eisenhower's New Look rearranged the financial and organizational priorities by which the Cold War was waged. Which is not to say it did not come to affect materials and bodies as well.

ARMIES AND ATOM BOMBS

In 1952, the year of Eisenhower's election, the stiffening fabric of the Cold War saw thousands of American soldiers mired on the Korean peninsula. Post-World War II reductions in defense spending (which dropped troops from 12.1 million to 1.6 million between 1945 and 1947, and the budget from $82 billion to $12.8 billion) had proved fleeting. By 1952 the Korean conflict had brought the army back up to 3.5 million troops. The defense budget climbed to $34 billion, fueled both by the war and by Harry Truman's policy (outlined in his report NSC/68 of 1947) requiring parity in spending with the Soviet Union.[2] And the Truman administration, which had been forcefully lobbied at its outset to "bring the boys home" and cut military spending, was suddenly defending itself against an opposite charge: of starving the military, and so being "soft" on the Communist threat. It was in this climate that Eisenhower was swept into office.

Dating from his first clerical duties in the 1930s, and increasing in subsequent responsibilities, Eisenhower's world view was one constantly filtered through the medium of typescript reports and executive summaries.[3] He was a marshal chiefly of information. Upon taking office as president, Eisenhower's first action was to centralize and organize the information on the ongoing Cold War flowing to the Oval Office. He chose as his vehicle a previously moribund body, the National Security Council (NSC).

Created in the same 1947 Act that established the Department of Defense, the NSC had been a peripheral body in the Truman administration; the president attended only twelve of the council's 57 meetings between 1947 and 1950.[4] Eisenhower moved the Council into the Executive Branch and re-created it as an advisory committee to the president.

Shortly thereafter, in an exercise code-named "Solarium," three separate NSC task forces were charged with preparing proposals for U.S. defense, based on the council's understanding of the choices available to them. These were (a) a continuation of the status quo; (b) the exclusive pursuit of nuclear capacities to intimidate the Soviet Union militarily; and (c) the idea of "liberation," or preemptive attack on the Soviet Union and its allies. After sixteen summer weeks in the Naval War College basement, the teams presented their proposals for a decision by the president.

At the July 30 meeting, however, Eisenhower was "put out."[5] The individual task force proposals were too divergent, and too nightmarish, to be understood as realistic alternatives. What resulted instead, and drew the exercise into the wintry fall of 1953, was an attempt at measured synthesis. The final outcome of the Solarium Exercise—NSC 162/2—contained three sections synthesizing each of the three committees' respective deliberations.

The first, contentious, point concerned finances, and was presented in language as guarded as the assets it discussed. While the report opened by stating its goal as "meet[ing] the Soviet threat to U.S. security … [without] seriously weakening the U.S. economy," opinions in the government were sharply split on how to shop for the necessary weapons. Eisenhower himself was the greatest proponent of fiscal conservatism.

Dwight Eisenhower had never so much as taken a mortgage prior to taking possession of the White House.[6] Against the advice of his own military establishment, he was adamant that deficit spending on the Korean War should not continue into a longer, colder conflict. The Joint Chiefs, by contrast, termed the economic threat of Cold War defense spending "no more than … incidental."[7] Only one of the three NSC teams, led by George Kennan, had openly worried about Cold War-related deficit spending on the American economy.[8] The final report recommended fiscal economy—but only within limits which the defense establishment itself set for an "overwhelming" response to perceived Soviet aggression.

The second issue considered by NSC 162/2 was the massive presence of U.S. troops in Europe and East Asia. Partly a relic of World War II, these troops were newly perceived as an essential guard against the armies of Russia and China. But their cost was enormous. Surprisingly, it was the hawks in the Solarium discussions who most clearly advocated pulling U.S. troops from foreign soil. In the face of what they considered an inevitable global nuclear war, the maintenance of an easily targeted offshore legion, they believed, consumed resources better spent on more atomic bombs.[9]

The final and most revolutionary section of the document, however, was its extensive discussion of nuclear weapons. During the debate over Korea, both Eisenhower and Secretary of State Dulles were alarmed at what they saw as a reluctance to consider the use of the most powerful and expensive weapons in the U.S. arsenal, agreeing in March 1953 that "somehow or other the tabu [sic] which surrounds the use of atomic weapons would have to be destroyed."[10]

While the different Solarium committees were essentially in agreement over the importance of an American nuclear arsenal to match and exceed the Soviets', resistance to nuclear spending came from an unlikely quarter—the Joint Chiefs of Staff themselves. Army Generals and Naval Admirals perceived nuclear capabilities (by then under control of the Air Force's Strategic Air Command) as drawing resources away from their traditional military domains. Yet the final language of the report was emphatic: "in the face of the Soviet threat," it concluded, the United States must have a "strong military posture, with emphasis on the capability of inflicting massive retaliatory damage by offensive striking power."[11]

In areas of funding, troop commitments, and nuclear strike capability, the New Look marked a seismic shift. The NSC recognized, at least in principle, that responding to the Cold War with an open checkbook could bankrupt the country just as much as open war. But the larger battlefield of the Cold War (versus individual conflicts like Korea) would become the nation's financial priority.

While the United States would retain troops in Europe and Korea until the close of the twentieth century, new foreign deployments were deemphasized in the 1950s in favor of a massive, nuclear arsenal. The technological nature of that arsenal was established as a continual effort at the state-of-the-art, planting the seed for the decades-long arms race to follow.

One month after Admiral Radford outlined the New Look in his Press Club speech, John Foster Dulles as Secretary of State addressed the Council of Foreign Relations in New York, framing "massive, retaliatory power"—nuclear weapons—as the central plank of U.S. foreign policy. Such a policy would reshape not only the size of the nuclear deterrent, but also its form, breaking the prior focus on strategic bombing in favor of a new delivery mechanism, the intercontinental ballistic missile (ICBM). Contemplated periodically ever since Wernher von Braun's successful missile attacks on London and Antwerp, the idea of an ICBM had not—until Eisenhower's New Look—been deemed a priority for the nation's defense. It would soon supplant all previous weapons in prominence, cost, and complexity.

MECHANICALLY MINDED FELLOWS

"I see a manless Air Force. ... For twenty years the Air Force was built around pilots, pilots, and more pilots. ... The next Air Force is going to be built around scientists—around mechanically minded fellows."[12]

Speaking to a panel studying the future of air warfare in 1945, General Henry (Hap) Arnold was the head of the largest manned air force ever established. Months before his retirement, though, Arnold saw several trends—automation in weaponry, the development of powerful nuclear devices, and the successful deployment of "manless" ballistic and cruise missiles by the German Luftwaffe—leading to a vision of future war in which a human airman was history.

Yet in the near-decade between the end of the war and NSC 162/2, Arnold's view was in the minority. Beyond the innate resistance of the "pilots, pilots, and more pilots" themselves—whose numbers constituted the new service branch's hierarchy—seemingly intractable problems in missile development slowed, and ultimately stalled, their progress until the mid-1950s.

The Air Force's first foray into ballistic missiles lasted from only 1946 to 1947. In seeking to develop missiles in the postwar era, the newly separate Air Force followed the same procurement procedure as the World War II-era army had before it—soliciting bids from military contractors, then delegating the construction of a working experimental ("X") prototype and follow-on trial ("Y") version, followed by the final product. Such contracts continued in the postwar era as the Air Force sought to replace World War II-era bomb delivery systems with planes more suited to nuclear weapons. (Of particular importance were new planes that could fly for long periods in the upper atmosphere, and fly long polar routes into Soviet territory.)

In their missile-development efforts, the Air Force lacked the in-house rocket expertise of the army, who retained Nazi rocket pioneer Wernher von Braun and his staff in their Huntsville, Alabama-based Ordnance Division. The Air Force contracted its rocket research to a favored outside contractor, the California-based Consolidated Vultee Aircraft Corporation, or Convair.[13] While instituting several improvements to the German V-2 design, Convair's few launch attempts of its new "Atlas" missile ended in spectacular failure, and the Air Force withdrew funding in 1947.

It was not until February of 1950, however, that anyone noticed. On Valentine's Day of that year, the *New York Times* reported that "a review of this country's guided missile program [is] urgently proposed … by Senator Lyndon B. Johnson, Democrat of Texas. The Russians, Johnson asserted, were ahead two years or more in that field. 'Our missile program is a minor, almost obscure item in the defense budget,' he added, 'Publicity about some of our rocket research has created a largely false impression that we have missiles which would be used in the defense of this country. As for now, we have none.'"[14]

While Lyndon Johnson's pronouncements on U.S. missile development added favorable publicity to the new senator's political profile, the resulting movement in the executive branch was mostly that of paper. With great fanfare, new studies of the potential for ballistic missile production were commissioned from Convair and the newly constituted air force consultancy RAND. These, however, largely concurred with the prospects assigned to ICBM construction by Vannevar Bush in 1945: "I say, technically, I don't think anybody in the world knows how to do such a thing, and I feel confident it will not be done for a very long time to come."[15]

NEW LOOKS IN ATOMIC TESTING 1952

Throughout the 1940s and 1950s, the US conducted a series of pivotal nuclear tests in the Pacific Ocean. Initially these tests were public, designed to display nuclear dominance to the world in the immediate aftermath of the Second World War. In the service of publicity for the first test, the 1947 Bikini Atoll explosion termed "Operation Crossroads,"[16] the joint military testing apparatus commandeered 1.5 million feet of motion picture stock and one million still negative frames, causing film shortages worldwide.[17] So much footage was exposed in this first Pacific test that the director Stanley Kubrick was able to produce the entire apocalypse concluding 1964's *Doctor Strangelove* from several seconds of a single Bikini explosion.[18]

After high-altitude sampling confirmed the successful test of a Soviet nuclear bomb in 1949, however, the Pacific nuclear tests were highly secret, focused on the development of the "Super" or hydrogen bomb, which would dwarf previous, uranium-based weapons. In 1951, the testing program had moved from the decimated Bikini Atoll to the Eniwetok Atoll, where on Elugelab Island, in the spring of 1952, U.S. scientists assembled a giant prototype H-bomb, code-named Mike. Resembling not so much a streamlined bomb as a "small oil refinery,"[19] the device was detonated two days before Eisenhower's election on October 31, 1952. Where the Baker blast had rendered Bikini uninhabitable, the 65-ton Mike removed the island of Elugelab from the surface of the earth, replacing it with a crater five times as wide as the Pentagon, and as deep as the Empire State Building.

The effect on public anxiety was tangible. When Edward Steichen collaborated with architect Paul Rudolph to design the famous exhibit "The Family of Man," they struggled over how, and whether, to integrate recent atomic developments. In the first version of the show, which opened at New York's MoMA in 1955, a Kodachrome transparency of the Mike blast was given its own, darkened gallery, the only color image in the show. Yet, while the exhibit as a whole toured the country, and later the world, the image of the bomb was withdrawn—too difficult to install was the official explanation; too incongruous, too disturbing, others admitted.[20]

Along with general anxiety, though, a subtler legacy of 1952's nuclear testing would spur the development of ICBMs. A series of much smaller tests in the Nevada desert the same year showed the potential for miniaturizing the new hydrogen bombs into much smaller sizes, drastically reducing the scale of the rockets needed to fling them around the curve of the planet. As with later developments in manned rockets, scientists quickly understood the cascading economies in a missile's size that came from a reduction of its payload.[21] When in 1953 it became clear that a thermonuclear weapon could shrink to less than one ton in weight, the ICBM theoretically required to carry such a weapon could shrink in size exponentially, bringing it close to the bounds of reality.

Less than a year after Mike, on August 12, 1953, the Soviet Union exploded its own hydrogen bomb. Evidence in the atmosphere indicated to the U.S. that the Soviet bomb was not nearly as powerful as the Mike blast. However, its design—deduced from traces of lithium in fallout—was smaller and more sophisticated than Mike's "refinery." To U.S. observers, it seemed that the Soviets had not only caught up with the United States in nuclear weaponry, but were possibly pulling ahead.

TEA GARDENS
A new Committee was formed. This one featured wartime computer expert John von Neumann as its chair, joined by physicist Edward Teller (whose Livermore National Laboratory had designed the thermonuclear weapons tested at Elugelab and in Nevada). The committee studied the potential effect of the new, lighter H-bombs on Air Force weapons development, and were unanimous in their recommendations toward missile development over reliance on traditional bombers and pilots.

And so, from the von Neumann committee's recommendation, an ambitious new Air Force assistant for research and development, Trevor Gardner, formed a top-secret group to evaluate—and potentially develop—ICBMs. Gardner asked that the committee, officially the "Strategic Missiles Evaluation Committee," be given a peaceful-seeming code name after his own initials; he proposed "Tea Garden." Air Force clerks, however, informed Gardner the name had already been taken, and assigned his work the even more prosaic "Teapot."

THE TEAPOT COMMITTEE 1953–1954
While spawned in the slow-moving bureaucracy of the Pentagon, the Teapot Committee drew upon Gardner's experience of the freewheeling interchange between military, industrial, and academic organizations in the desert climate of southern California. There, in a short-sleeves climate, and shirtsleeves informality, Gardner had spent the years between 1939 and 1948. He had floated freely between loosely connected institutions, first as a military liaison at Caltech, then executive at General Tire and Rubber, and finally as a founder of an electronics startup, Hycon Industries, which he left for the Department of Defense in 1949.[22] The growing complexity of aviation technology had led to collaborative connections between private business, technical academia, and public military enterprise, With the free exchange of technical managers such as Gardner, these connections were further reinforced. The resulting network produced some of the most technically advanced World War II hardware, especially in the growing field of airborne electronics. Gardner, and men like him (including the Teapot Committee's technical liaison, engineer and Air Force

4.2 (preceding pages)

Reconnaissance personnel pose with
cameras to be loaded into their observation
plane, Bikini tests, 1947.

4.3

View of the Ivy Mike blast.

officer Bernard Schriever), thrived in such technically demarcated informality, and so—perhaps inexorably—advocated its re-creation at larger and larger scales.

In his own ICBM committee, Gardner selected a stable of thoroughbred advisors from military, industrial, and academic backgrounds. These included John von Neumann, MIT President Jerome Wiesner, and executives of Hughes Aircraft, Bendix Aviation, and Bell Laboratories. Simon Ramo and Dean Wooldridge served both as committee members and research staff; they were principals of a new firm that, with Charles Thornton's Litton Industries (see Layer 17), had split from Hughes Aircraft's Electronics Division in 1952.

The committee's final report, delivered in February of 1954 (just weeks after Dulles's "massive retaliation" speech), substituted at the last minute its assertion that "the Russians are probably significantly ahead of us in long-range ballistic missiles" for the more moderated suggestion that such an idea "could not be ruled out."[23] When combined with the parallel strategic shifts of the New Look, however, the Teapot Committee report—and the host of presentations its members gave to individual decision-makers—produced an enormous shift in the Air Force and government's spending priorities.

In the strategic framework of Eisenhower's New Look (in which the United States would best the Soviet Union with retaliatory nuclear technology and not by strength of ground forces), the committee's assertion that the United States could be "behind" in such a rapidly developing area of nuclear technology could not be easily suppressed. On May 14, 1954, the Army made a revived Convair Atlas program its first priority for research and development funding. By December 8, 1955, the same National Security Council that had overseen the birth of the New Look in defense was recommending to the president that the ICBM program should be given "highest priority over all others."[24]

SCHRIEVER, RAMO, WDD, INGLEWOOD

Through 1954 and 1955, Trevor Gardner found his time filled with presentations, consultations, and exhortations—increasingly focused on the perceived "missile gap" with the Soviet Union. In his frequent absence, Simon Ramo, with assistance from Bernard Schriever, chaired the missile group. As the committee was charged with transforming itself into supervising the accelerated production and deployment of strategic missiles, Lieutenant General Schriever was also named commander of a newly constituted "Western Development Division (WDD)" of the Air Force's research arm. Its first (secret) headquarters was an abandoned schoolhouse on the outskirts of Los Angeles International Airport in Inglewood, California.

RADICAL REORGANIZATION

Instead of simply reactivating the existing Convair Atlas contract, the Teapot Committee had recommended a "radical reorganization." Instead of a single contract with a private firm, ICBMs would be developed using a design "system." Instead of building the missiles themselves, the Ramo-Wooldridge corporation would be contracted to supervise and integrate the work of a range of public and private entities centered at the WDD. As a result, Simon Ramo and Dean Wooldridge recused themselves from the committee but became even more enmeshed in the project's complex organization.

At the head of the ICBM project, Ramo-Wooldridge (R-W) became a new kind of military contractor, a "systems engineer." Producing no physical artifacts, R-W instead developed the systems by which those artifacts would be designed. As well as the immediate needs of the Atlas ICBM, R-W's contract made explicit reference to what the Teapot Committee report termed the "advanced research phases of a continuing [ICBM] program." In the reality of the Cold War arms race acknowledged and amplified by the New Look, military hardware development, or design, was no longer an X, Y, or Z product to be designated but, instead, a perpetual process, of which continuously improved weapons were anticipated to be only one tangible result.

The notion of a "systems engineer" both accentuated, and flattened, the multiple levels of "system" involved in the design of the ICBM as an object. After setting up its own administrative organization, or system, to work with Schriever's WDD, R-W was charged with crafting a strategy as to what kinds of systems—guidance, propulsion, fueling, structure, "warhead"—the missile would have to contain. As well as delegating the development of these "subsystems" to subcontractors, Ramo-Wooldridge would have to specify exactly which substance the systems conveyed to each other—fuel, thrust, guidance information, structural support, and so on. While overseeing contractor research and development on each subsystem to ensure their eventual integration, the firm would ultimately be responsible for testing these systems separately, and then together, to ensure that the final object functioned as expected (or, more frequently, to determine why unexpected results appeared). These results would be delivered, along with their budgetary ramifications, back into the Air Force administrative "system" that supervised and funded the missile effort as a whole.

The physical prototypes that resulted from the swiftly accelerating Atlas efforts were not only products, or even symptoms, of this continuous "systems engineering," but complex systems in themselves, connecting and blurring information, vectors, fluids, and mechanisms in a self-guided, self-detonating death machine independent (once targeted and launched) of human intervention.

BLACK SUITS AND BLACK BOXES

While interdependent, however, these proliferating systems were not transparent. Indeed, the systems of the missile, and the system of production that surrounded it, were self-consciously disguised and impenetrable. The military personnel working side by side with Ramo-Wooldridge engineers in the growing Inglewood complex shed their uniforms for standard office dress. This was ostensibly to improve security at the guarded facility, but also to encourage better integration of military officers with their civilian colleagues. When teams of engineers from subcontractors gathered to assemble the missile system in prototype form, individual components were boxed, often in black, emphasizing and revealing only their connections to each other. As a result, the idea of a "black box" came to stand in for the notion that one did not need to know how a subsystem or component worked, only its characteristics as it connected to other parts of the whole.

"LIKE NONE EVER BUILT BEFORE"

When, on April 29, 1957, *Time* magazine devoted its cover story to the emerging "systems industry," "Engineers Wooldridge & Ramo" (in contrasting brown and blue suits) were chosen as the faces of the new field, and their ICBM-creating system was reported as the "giant" of the new age.

"The U.S. military establishment," *Time* reported, "is rapidly becoming one vast system." Lest a newsstand patron scanning the cover miss the significance of the unremarkable Ramo and Wooldridge in business attire, looming behind them in ensemble was a giant technological golem, with a radar screen and oscilloscope eyes, around which, in the far distance, zoomed the planes and missile (one hoped) securely in its control.

The blurring between invisible technological systems and physical objects was so unusual to the author of *Time*'s Ramo-Wooldridge profile that, instead of explaining the ICBM system's global reach, they instead commenced with a conjecture as to the domestic implications of the systems engineering approach.

"The house was like none ever built before," the profile of Ramo and Wooldridge began. "Doors and windows opened in response to hand signals; they closed automatically when it rained … at night, walls and ceilings glowed softly … [and] changed color at the twirl of a dial. … Such a house," *Time* asserted, "is today a reality in the laboratories that are moving deeply into the coming age."

Yet almost nothing of the ICBM project was domestic in scale. In 1957 the effort included 75,000 workers in hundreds of congressional districts, and would result not only in the multistory Atlas missile but a massive "backup" system, the Titan. Once missiles became operational in 1958 and 1959, the reach of the project became truly global, blurring the line between the physical objects of the ICBM program and the information that alternately described and determined its development.

After the Soviets' 1957 orbital launch of Sputnik (which, not incidentally, displayed an ability to place a ballistic warhead anywhere in the United States), pressure on Schriever's team only increased. At the same time, Schriever and Ramo-Wooldridge's ICBM program met several obstacles in their goal of "systems engineering." In each case, these hurdles seemed to initially represent a failure of the approach, but ultimately led to its extension and expansion into further realms of physical and informational reality. These obstacles had first to do with the internal funding that flowed through procurement contracts, and secondly with the interface between the different components produced by over 220 contracts in the missile effort.

While before the Second World War contracts for military construction resembled those in the private sector, with contractors bidding on the basis of both cost and perceived quality, a war-fearing Congress in 1940 had legalized a new kind of contract, in which a contractor was paid a fixed fee for producing a plane or tank—but passed on all the "costs" of producing the weapon to the government as well. Before the temporary measure expired in peacetime, the military and industry successfully lobbied for a 1947 Act of Congress that allowed the process to extend into peacetime, and thus the Cold War.[25]

In the lull before Sputnik's launch, the same congressional committees that complained about a perceived "missile gap" were now complaining about the spiraling cost of missiles to the American public, and the lack of financial accountability in the systems engineering approach. This problem was linked in both substance and, ultimately, solution to another set of difficulties encountered in the myriad web of contractors supervised by Ramo, Wooldridge, and Schriever. Sometimes, problems were tangible; as components were separately improved, they would no longer fit. Sometimes, they were distinctly immaterial; a shifting frequency in a pump in the

4.4
An Atlas missile exploding on launch.

missile's base would play magnetic havoc with a spinning gyroscope in its nose. Even more subtly, as missile prototypes began to be fired, the destruction of the missile by the missile test, whether successful or not, meant that it could often not be determined exactly what sort of object, tuned and tweaked to the last minute by subcontractors and R-W engineers, had succeeded or failed.

In efforts to manage the physical fabric of the machine as well as its cost, R-W developed a system of "configuration management" where first the form, then the estimated cost, of individual versions of missile components were fixed through a system of memoranda flowing to and from Western Development Division offices. To its subcontractors, Ramo-Wooldridge became known as "the paper factory," producing reams of assessments and directives to manage the substance and expenditure of the missile program.

By the end of the decade, the process first developed in Schriever and Ramo-Wooldridge's California offices to meet the demands of systems engineering surfed a wave of technological transformation throughout the entire military. Nuclear submarines, jet aircraft, and battlefield ordnance all became subject to analogous or identical processes. These formed a supersystem of sorts in which systems engineering methods were tested and extended. A further addition in the early 1960s was "systems requirement analysis,"[26] in which the performance requirements of a specific object or subsystem were described separately from its configuration for budget and engineering purposes, codifying what it did as well as how it did it, and making the paper life of the object from conception to execution as concrete, if not more so, than its physical substance.

THE MILITARY-INDUSTRIAL COMPLEX

As the physical complex shared by Ramo-Wooldridge and Schriever's Western Development Division in Inglewood expanded, it both reflected and shaped the organizational reality it embodied. From an incognito schoolhouse it moved to a pink and green purpose-built complex in 1955. Inside the office, military officers in business suits worked in spatially mirrored cubicles with their Ramo-Wooldridge counterparts (academic advisors and subcontractors occupied outer circles). At the administrative center, Bernard Schriever and Simon Ramo occupied adjacent offices and met almost every day, as did their respective deputies.

Bernard Schriever had originally planned to be an architect, and had only been driven into the military by the Great Depression. But one space he designed for the Inglewood Complex would become as influential in the architecture of the century itself as the work of any émigré modernist. At the center of the complex (and indeed of the entire ICBM effort), housed in a protective concrete bunker and constantly staffed, was the "project control room." Here, Schriever and trusted lieutenants—whether commissioned officers or Ramo-Wooldridge managers—surveyed the scope of the entire development system, first with paper charts, then, increasingly, with new digital computer printouts and eventually projections.

Not a house of tomorrow, the control room was nevertheless inhabited by what *Time* magazine (in a 1957 profile of Schriever) termed "tomorrow's men." *Time* described the technologically enhanced decision-makers as "ebullient scientist[s] … fresh-faced lieutenant[s] from MIT handling millions of dollars worth of rocketry … or a gentle [sic] German in tweeds who helped Hitler."

"Tomorrow's man" himself was Bernard Schriever, "run[ning] his command post in a grey flannel suit," blurring the lines between corporate, military, and academic organization just as surely as "the paper factory" itself blurred the line between system and object, between administrative influence and actual instance. By 1959 the Inglewood opaque control room, guarded by defenses so sensitive they could "register an intruder's breath,"[27] stretched its administrative influence over a network larger and more complex than the Manhattan project, and commanded funds amounting to nearly 10 percent of the entire Federal Budget. What started as the "project control room" soon became known simply as the "control room."

Developing alongside the operational control rooms of America's nuclear defense system (see Layer 18), the Inglewood Control Room underscores how the development of continuously more complex technologies paralleled, and even superseded, the more explicit task of military defense. The control room, an architectural typology originally derived from the moment-to-moment need for information in warfare, became in postwar defense both a method for coordinating computerized simulations but also, in Schriever's hands, a method for coordinating an even more challenging task: constantly administering continuous improvements in defense technologies. The first of its type,[28] the Inglewood Control Room was a meta-environment for the control of military technology, revealing and shaping a set of continuously changing systems, which themselves would be subject to further spaces of control in their subsequent deployment.

LOOKING AT THE NEW LOOK

Eisenhower's New Look of 1953–1954 did not explicitly forecast this architecture—a military-industrial "complex" blurring the lines between corporations and defense commanders, continuously expanding its scope with the onrushing equivalencies of the Cold War arms race. But it was the rapid and likely inevitable result. The dismantling and reassembly of previously separate organizations—the army, the company, and the university—was seen as the only way to integrate the correspondingly complex and separate systems of ICBM development. That such an organizational synthesis had the power to innovate and accelerate military technology was clear. The potential drawbacks were also perceived from the outset. (When, during the Solarium discussions, President Eisenhower had contemplated a proposed military takeover of American industry, he snapped: "Well yes, we could lick the whole world if we adopt the system of … Hitler!")[29]

As the system to which the New Look gave birth expanded through the remainder of his presidency, Eisenhower's concerns grew as well. His farewell address to the nation, delivered via national television broadcast on April 17, 1961, was substantially a warning to the country about the Pandora's box Eisenhower knew he had opened:

> We have been compelled to create a permanent armaments industry of vast proportions. Added to this, three and a half million men and women are directly engaged in the defense establishment. We annually spend on military security alone more than the net income of all United States corporations.
>
> Now this conjunction of an immense military establishment and a large arms industry is new in the American experience. The total influence—economic, political, even spiritual—

4.5
A Convair Atlas (SM-65) guided missile
deployed in concrete-hardened silo,
South Dakota, 1963. The same craft would
loft Mercury astronauts into orbit.

is felt in every city, every Statehouse, every office of the Federal government. We recognize the imperative need for this development. Yet, we must not fail to comprehend its grave implications. Our toil, resources, and livelihood are all involved. So is the very structure of our society.

In the councils of government, we must guard against the acquisition of unwarranted influence, whether sought or unsought, by the military-industrial complex. The potential for the disastrous rise of misplaced power exists and will persist. We must never let the weight of this combination endanger our liberties or democratic processes. We should take nothing for granted.[30]

Notes

1. Saki Dockrill, *Eisenhower's New-Look National Security Policy* (New York: St. Martin's Press, 1996), 54.

2. Ibid., Appendix 1, "U.S. Military Personnel Strength, 1945–60," 281; and table 3.1, "Outlays by superfunction and function: 1940–2009," in Office of Management and Budget, *Historical Tables, Budget of the United States Government, Fiscal Year 2005* (Washington, D.C., 2004), 45.

3. For a discussion of Eisenhower's bureaucratic penchant and skills, see Stephen E. Ambrose, *The Supreme Commander: The War Years of General Dwight D. Eisenhower* (Garden City: Doubleday, 1970). Ambrose credits Eisenhower's D-day bureaucracy as being "much more scientific than anything Clausewitz ever imagined possible," 471.

4. Alfred D. Sander, "Truman and the National Security Council: 1945–1947," *Journal of American History* 59, no. 2 (September 1972): 387.

5. "Cutler Minute, 16 July 1953," *Foreign Relations of the United States 1952–4*, National Security Affairs, vol. 2, 397–398.

6. Eisenhower had even turned down what he thought in 1946 would be his retirement home from the army in San Antonio, "confessing that, never having had a Mortgage before, he was highly uncomfortable at the thought of committing himself to payments for the next 12 years." Stephen E. Ambrose, *Eisenhower: Soldier and President* (New York: Touchstone/Simon and Schuster, 1990), 234; and from his farewell address: "We cannot mortgage the material assets of our grandchildren without risking the loss also of their political and spiritual heritage."

7. "Cutler Minute," *Foreign Relations of the United States 1952–4*, 2:523.

8. Instead of revealing these strains, the final report masked them behind language that, while advocating a "strong, healthy and expanding U.S. economy," also allowed that the ultimate cost of "necessary" security policies "cannot be estimated until further study."

9. As with the more general question of financing the Cold War, disagreements over America's offshore commitments were finessed rather than finalized by NSC 162/2's text. While redeployment of these troops was mooted, the notion that such a strategy should be engaged "reasonably soon" was deleted, leaving the de facto commitment of resources unchanged. Dockrill, *Eisenhower's New-Look National Security Policy*, 39.

10. "Cutler Minute," *Foreign Relations of the United States 1952–4*, 15:827.

11. Against this language, the Joint Chiefs were only able to introduce a preamble to NSC 162/2, reserving the right to petition to change the word "emphasis" to "include" if they felt the report's

strategy "placed the National Security of the United States in jeopardy," Dockrill, *Eisenhower's New-Look National Security Policy*, 42.

12. Theodore von Kármán, *The Wind and Beyond* (Boston: Little, Brown, 1967), 271.

13. Convair's giant B-36 bomber was central to nuclear bomb delivery throughout the 1940s and 1950s, and was featured, piloted by Jimmy Stewart, in the 1959 film *Strategic Air Command*.

14. William S. White, "U.S. Willing to Discuss Atom with Soviet Union in U.N." *New York Times*, February 14, 1950, 1.

15. Thomas Parke Hughes, *Rescuing Prometheus* (New York: Pantheon Books, 1998), 79, note 29.

16. Four days after the first Crossroads blast, Louis Réard, a 47-year-old engineer attempting to rejuvenate his family's lingerie business, held a press event at the Molitor hotel in Paris. The navel-baring two-piece swimsuit he featured was not new—a recent version, named the "Atom" for its tiny size, had been marketed in Cannes that spring—but its name was: "Bikini." Kelly Killoren Bensimon, *The Bikini Book* (New York: Assouline, 2006), 12.

17. Michael Light, *100 Suns: 1945–1962* (New York: Alfred A. Knopf, 2003), 58.

18. Jonathan M. Weisgall, "The Able-Baker-Where's Charlie Follies," *Bulletin of the Atomic Scientists*, May/June 1994, 26.

19. Richard L. Miller, *Under the Cloud: The Decades of Nuclear Testing* (New York: Free Press, 1986), 116.

20. John Szarkowski, "The Family of Man," *The Museum of Modern Art at Mid-Century: At Home and Abroad* in *Studies in Modern Art*, vol. 4 (New York: Museum of Modern Art, 1994), 22.

21. A pound of payload at the tip of a rocket is lifted by hundreds of pounds of fuel in the first stage, which translates to thousands of pounds to lift those hundreds in the second stage, etc.

22. Jacob Neufeld, "Technology Push," paper given to the Colloquium on Naval History, Naval Historical Center, Washington, D.C., September 2003, <www.history.navy.mil/colloquia/cch9c.html> (last accessed March 19, 2008).

23. "Recommendations of the Tea Pot Committee," February 1, 1954, in Jacob Neufeld, *The Development of Ballistic Missiles in the United States Air Force, 1945–1960* (Washington, D.C.: Office of Air Force History, U.S. Air Force, 1990), 260–261.

24. Dockrill, *Eisenhower's New-Look National Security Policy*, 178.

25. Stephen B. Johnson, *The Secret of Apollo, Systems Management in European and American Space Programs I*, New Series in NASA History (Baltimore: Johns Hopkins University Press, 2002), 23.

26. Ibid, 63.

27. "The Bird and the Watcher," *Time*, April 1, 1957, 8.

28. Hughes, *Rescuing Prometheus*, 125–126.

29. Dockrill, *Eisenhower's New-Look National Security Policy*, 36, note 116.

30. Farewell address by President Dwight D. Eisenhower, January 17, 1961. Final TV Talk 1/17/61 (1), Box 38, Speech Series, Papers of Dwight D. Eisenhower as President, 1953–61, Eisenhower Library, National Archives and Records Administration.

Layer 5   FLIGHT AND SUITS

5.1
Wiley Post and his pressure suit in Columbia
Pictures' *Air Hawks*.

"AIR HAWKS" AND HEROES

The first pressure suit constructed in the United States was also the first featured on film. When the celebrated pilot Wiley Post strode onto the screen of 1935's science-fiction adventure *Air Hawks*, he was wearing a silver-helmeted pressure suit which—the script assured us—would help Post carry mail far above the reach of the evil Professor Schulter's "electric death ray."

Death ray aside, Post's suit was the genuine article; it would carry the flier into the stratosphere dozens of times in 1934 and 1935, years before the U.S. Army Air Forces would venture regularly to such heights. Also genuine was the publicity that surrounded the pressure suit's debut (it appeared in *Air Hawks* only weeks after it was manufactured). The exposure highlights the dense network of media, perception, and human performance through which, and for which, pressure suits would thereafter be designed.

THE GLOBE AND THE ZEPPELIN

When Wiley Post set out to capture his first aerial record of the 1930s, it was to show the world that the future of flight lay in airplanes, as opposed to the massive, rigid dirigibles that were then the icons of modern flight.[1] An apostle of the sky, Post deliberately searched for an achievement that would convince the media, and thus the public, of the safety and speed of powered flight for global travel. As the mighty stable airships of the 1920s crossed the Atlantic and the globe, the primacy of a tiny, buzzing aircraft was hardly assured.

The large dirigible (literally, "steerable") balloons of the 1920s and 1930s were known colloquially by the title of their most famous engineer, the German Graf (Count) Ferdinand von Zeppelin. While the Count himself traced his airship's origins to a visit to the United States in the 1860s,[2] he would not fly his first successful airship until 1900, when the LZ1 rose above Lake Constance in present-day Austria. The Count lived to see more than 80 of his ships participate in the First World War, flying nonstop as far as Khartoum and the Arctic to ferry troops, drop ordnance, and make observations. These exploits fulfilled the Count's vision of the airship as instrumental in modern warfare, first recorded in his diary during the Franco-Prussian War of 1870–1871.

Zeppelin's death in 1917 spared him the prospect of German defeat. After the armistice, it seemed as if the provisions of the Treaty of Versailles would shut down the Luftschiffbau Zeppelin AG permanently. Salvation arrived in 1921 as the U.S. Navy ordered the first of a new generation of airships (ignoring European protests). Over 700 feet in length, carrying a passenger load of 20, the 56-hour delivery flight of the silver United States Airship *Los Angeles* from Frankfurt to Lakehurst, New Jersey, in 1924 made headlines the world over.

Allied restrictions on German civil aviation were lifted in 1926, and in 1928 Zeppelin AG christened the LZ127 *Graf Zeppelin*, in its founder's honor. Shattering the record of 175 days for a round-the-world aerial voyage set by a U.S. Army Air Service expedition in 1924 (the team made 69 stops), the *Graf Zeppelin* accomplished the same feat in 21 days in August 1929. It carried a full load of passengers and made only three stops—Germany, Japan, and Los Angeles. The flight was financed by newspaper magnate William Randolph Hearst, who had insisted that the journey begin and end at the Statue of Liberty. By 1931, the *Graf Zeppelin* offered a regular passenger service from Frankfurt, via New Jersey, to Rio de Janeiro.

CELEBRITY AND CIRCUMNAVIGATION

It was to beat the enormous airship's 21-day circumnavigation that pilot Wiley Post set off in 1931, in a tiny Lockheed Vega named the *Winnie Mae* (after the teenage daughter of his benefactor, Oklahoma oilman F. C. Hall). "What I was attempting to prove," Post wrote after the flight, "was that a good airplane with average equipment and careful flying could outdo the 'Graf Zeppelin,' or any other similar aircraft, at every turn on a flight around the world."[3] The contrast between the two vehicles could not have been more extreme. While the Zeppelin's opulent interiors contained staterooms and lounges (with walls of fabric and latex foam to save weight), Post sat astride the crankcase of the Pratt & Whitney "Wasp" engine that powered his 1928 monoplane, his rear end wedged against a fuel tank. Yet Post (aided by Tasmanian navigator Harold Gatty) shattered the Zeppelin's record by two weeks, landing at Mineola field on Long Island seven days and eighteen hours after takeoff.

Post and Gatty returned to a welcome the *New York Times* compared "only to that given Colonel Charles A. Lindbergh on his return from Paris." The ticker tape along their Broadway parade route was so dense that Post, Gatty, and their wives repeatedly had to be extricated from the cloying streamers in order to wave their arms at the crowds.[4] Two years later, Post set out in the *Winnie Mae* once again to fly around the world. This time he was accompanied not by a human navigator

5.2
A business-suited Wiley Post in front of the
Winnie Mae, 1932.

but by "Mechanical Mike"—his name for the "robot" gyroscopic compass and autopilot that supplied him with navigational data. When the solo navigation was also a success, further acclaim established Post as a bona fide hero in the public imagination.

Post may have been compared to a machine for his dependable performance as a pilot, but an essential component of his success was a singular attention to his own physiological limitations. Unlike the parsimonious Lindbergh (who sat on a wicker stool), Post installed a stuffed club chair in the cockpit for his second round-the-world flight, declaring that "if he was going to sit by himself, he would do it well." He pioneered his own meditation techniques for maintaining alertness, striving to keep his mind as "blank as possible" for hours at a time. In his account of his original circumnavigation, Post was the first to discuss the effect of shifting time zones on human physiology—what we now know as jet lag.[5] A student of his own body, he attributed the strength of his flying vision to having only one eye, a result of the same mining accident whose insurance payout helped him purchase his first airplane.

A SUIT FOR THE STRATOSPHERE

When he flew around the world alone in 1933, Wiley Post was dressed in a dark gray double-breasted suit, worn over a white shirt with a blue necktie.[6] Unlike his barnstorming compatriots, who would often don cavalry boots and breeches to emphasize the adventurous exoticism of the air, Post deliberately wore street clothes to fly in, projecting a routine image onto the spectacular exoticism of flight.

In his solo circumnavigation, Post had several times been forced to climb above weather, beyond the ceiling of his own tolerance for altitude. He would remain at 15,000 and 20,000 feet for as long as he dared—up to several minutes—before the effects of low pressure set in. Already used to considering himself a vital and adaptable part of the plane's mechanism, Post saw in these brief glimpses of altitude the potential for smoother air and, he believed, fast-moving air currents that could aid in the prospect of rapid, reliable airplane travel. The limit to flying longer and further at these heights was not so much a problem for Post's plane or its engine—into which a "supercharger" could force air at an increased rate to support combustion—but for his own physiology. Even in his brief 1933 climbs, Post experienced symptoms of altitude illness—bleeding from the eyes and nose, discomfort, and headaches—which, to the cautious and systematic flier, seemed prohibitive.

By 1933, the technology of pressurized enclosures—earthborne altitude chambers, bathyspheres, and the stratospheric gondolas of balloons—was well developed. The delicate bodies of the era's airplanes, however—and especially the aging, spruce-shelled *Winnie Mae*—were not compatible with the heavy construction and machinery needed for a pressurized capsule. Instead of considering a pressurized airplane, Post was among the first to imagine a pressurized suit of clothing, allowing the *Winnie Mae*, or any plane, to be flown far above the conventional limits of human endurance.

Post's interest in such a device was practical. Neither he nor his patrons could easily afford a new plane, and for the upcoming 1934 MacRobertson Trans-Pacific race (which offered a $50,000 prize), it was far easier—Post concluded—to increase his own performance at altitude instead.

5.3
The first pressure suit designed by Russell Colley
and Wiley Post, 1932.

Post turned to friends in industry—in particular, World War I hero Jimmy Doolittle, then a Shell Oil executive—for suggestions as to who might build such a "suit." Doolittle in turn referred Post to the Los Angeles branch of the B. F. Goodrich Tire and Rubber Company, with which he had recently become acquainted. Post visited the plant in April of 1934 and requested "A rubber suit which will enable me to operate and live in an atmosphere of approximately twelve pounds absolute (5,500 feet altitude equivalent)."[7]

In the summer of 1934 the first prototype of the suit was tested at Wright Field in Ohio; at this point its development had been delegated to Russell Colley, an engineer at Goodrich's Akron plant. Colley was familiar with rubber as well as airplane flight from his pioneering work on de-icing mechanisms, vibrating "boots" of inflated rubber that broke off freezing water from a wing's leading edge. With Colley, Post attempted three suits and two helmet designs before a functional prototype was devised.

For the first suit, Goodrich engineers had turned to diving-suit technology, which seemed to offer the closest analog to Post's request. A fundamental difference between depth and altitude, however, sunk this first attempt. A diving suit is designed to protect the body against the exterior force of water. As a result, the pneumatic pressure of the suit's inflation is secondary to the physics of the surrounding fluid—and thus irrelevant to the wearer's mobility (save for in the deepest oceans). A suit for altitude, however, is inflated to a much higher pressure than the surrounding air. Its physics, then, have much more to do with a basketball than a bathysphere—and the first Goodrich suit reduced Post to an immobile, inflated statue.

For the next attempt, Colley and his team in Goodrich's Ohio plant focused on fitting the rubberized fabric as closely as possible to Post's body. This suit fit so well, however, that Post had to be cut out of it while standing in a golf ball storage refrigerator to avoid overheating.

With the third, successful model of the suit, the functions of pressure bladder and restraint were separated into a latex bladder and a bias-cut, canvas restraint layer designed to keep the latex bladder from ballooning. The latex pressure bladder was formed on a cast of Post's seated body, and the restraint layer was likewise tailored to his piloting form (with the result that Post could only walk hunched over to and from the airplane).

The final advance of this last suit was the transfer of its air supply from the supercharger of the airplane (which compressed atmospheric air for high-altitude combustion) to a dedicated cylinder of liquid oxygen. This allowed the suit to be inflated to a lower pressure while still providing the same volume of oxygen to Post's blood. Expanding from the compressed cylinder, the oxygen was forced through a hose into the side of the screw-on porthole of Post's new bucket-shaped helmet. On account of its owner's missing eye, the helmet's porthole was off-center, allowing the hose to deliver the frigid oxygen directly over Post's eye patch. The flow would have frozen a real or glass eye in the course of a flight, but the patch provided the perfect surface to warm the current and deflect its flow onto the porthole surface, dissolving any condensation before being breathed by Post and exhausted through a relief valve on the opposite side of the helmet.

The third suit was pressure-tested in August 1934. Additional layers of canvas, on differing biases, were added to control ballooning, and at the end of the month Post made the first altitude

5.4
Wiley Post boards the *Winnie Mae* for an altitude attempt.

tests of the mechanism in the Wright Field pressure chamber. In September, he made the first publicized flight using the suit, soaring to 40,000 feet above the Chicago World's Fair. The suit worked as predicted, although its extremely limited mobility meant that an extended handle was needed on the *Winnie Mae*'s joystick. The time taken for further adjustments to the engine and landing gear for high-altitude flight meant that Post missed the takeoff for the MacRoberson race in October. Instead, he turned his attention to setting new records for high-altitude flight.

When Post started his stratospheric expeditions in late 1934, the world altitude record of 44,352.219 feet was held by an Italian—Renato Donati. His brief aerial trip, however, like that of most of Post's competitors, was as much a product of pluck as skill; Donati exited his airplane in an ambulance and took several weeks to recover from the flight. By December 1934, by contrast, Post was regularly traveling to 40,000, 45,000, and 50,000 feet in order to test the engine and the pressure suit, with no ill effect, all as preparation for attempting a new transcontinental record.

Although meteorologists had noted upper-atmosphere clouds reaching ground speeds of hundreds of miles an hour in the nineteenth century, Wiley Post was the first to predict that such currents, and in particular the west-to-east "jet stream," could be used by aviators to achieve record speeds. It was with such a goal in mind that Post moved to Los Angeles in January 1935, preparing for what would become several eastbound speed attempts.

While the flights were plagued by equipment failure and even sabotage, Post would fly in March 1935 as far as Cleveland, in the record time of seven hours and nine minutes. At times, his ground speed was as high as 340 miles an hour, doubling the Vega's usual pace. Yet the mechanical limitations of the seven-year-old *Winnie Mae* were becoming increasingly apparent, and as a result Post interrupted his experiments and retired the plane. In the midst of attempts to secure a new plane (his adventures had long outstripped the resources of his early patron F. C. Hall), Post traveled to Alaska to vacation with a friend of equal celebrity, newspaper columnist Will Rogers. There, on August 15, 1935, he and Post were both killed while taking off in a pontoon seaplane from remote Point Barrow, when an engine failure forced the nose-heavy seaplane into the water at high speed.[8]

JET MEN AND MOON MEN

An enduring legacy of a too-brief career, Post's pneumatic suit was not designed to be an outer layer but rather an interface, or media, allowing pilot and plane to continue to interact beyond normal limits of human endurance. When Post was most successful—as when integrating himself with devices like his "Mechanical Mike" copilot of 1931—it was because, both through a mastery of his plane and his own physiology, he appeared to erase the boundary between man and machine. In his discovery and mastery of phenomena as diverse as jet lag and jet stream, the half-blind flier would seem in the end to have integrated his body and plane with the scale of the planet itself.

Yet Post's pressure suit was not designed to be exotic or spectacular. Like his previous flying garb of suit and tie, it was designed to help render flight a more pedestrian and everyday experience—although in this case by scientifically conveying Post's body to explore the stable and speedy stratosphere. After defeating the Zeppelin in his round-the-world flight, Post sought to create a pneumatic atmosphere for his own body, an anthropomorphic earthly atmosphere to take to the upper limits of the sky. Yet this was not to be the suit's final legacy.

While Post's larger exploits did much to create the dreary reality of today's commercial flight, his pressure suit would have more uncommon legacies. In technological terms, his work with B. F. Goodrich, while mothballed after his death, would emerge again in the Second World War as armaments, not adventure, drove pilots higher into the sky. But Post's suit leaves a further trace.

Like the plot of *Air Hawks* itself—valiant U.S. airmail pilots battling an electric death ray—Post's suit, and its celluloid appearance, fashioned together the factual and (science-) fictional. The opposite of Post's 1931 business suit—a self-consciously ordinary surface clothing an extraordinary man—the 1934–1935 stratosphere suit appeared on film over Post's (emphatically) ordinary body, as not just a portable environment but a symbolic and transformative attire.

In the opening credits of *Air Hawks*, when Post's name is flashed on the screen, it is superimposed on footage of the flier, pressure suit and all. And so, as in the cartoon universe of the day, the pressure suit did not hide the body of the aviator, but instead, by wrapping it in shining fabric and aluminum, revealed the futuristic fantasy with which his body became imbued. In the almost instant appropriation of his pressure-suited form by hungry media, Post became not only a pioneer of the jet age, but a pioneering version of the cultural type identified decades later by French critic Roland Barthes as the "jet man." He was "defined less by his courage than his [control of] habits ... [and an] apartness read in his morphology; the ... suit of inflatable nylon, the shiny helmet ... a novel type of skin. ..."

Barthes's 1955 vision was dystopic but accurate. The individuality of heroes like Saint-Exupéry and Lindbergh contrasted sharply with a new "fictitious, celestial race," depersonalized by their equipment and thus appearing to be "a kind of anthropological compromise between humans and Martians."[9]

As America's most prominent media outlet, the text-heavy *New York Times* did not illustrate its articles of Post's exploits with photos of Post or his clothing, but described in breathless detail the consequences of one of his engine failures, which forced him to land in the desert near Muroc, California, in February 1935:

> Graceful as a seagull, the Winnie Mae quietly dipped to the desert lake and slid along ... the only man nearby did not see the landing. He was a motorist, four hundred yards distant, tinkering with a balky auto engine.
>
> Post, attired in his grotesque stratosphere flying suit with cylindrical helmet, climbed out of his ship and walked to the stalled motor car. He tapped the motorist on the back.
>
> "The man's knees buckled and he almost fell over," said Post, describing the incident. "The sight of me in this pressure suit with oxygen helmet was a little too much for his heart. He ran around to the back of his auto and peered at me."
>
> Finally, the words of Post restored the man's courage.
>
> "Gosh, fellow!" he exclaimed when he found his voice, "I was frightened stiff. I thought you dropped out of the moon."[10]

Notes

1.   Bryan B. Sterling and Frances N. Sterling, *Forgotten Eagle: Wiley Post, America's Heroic Aviation Pioneer* (New York: Carroll & Graf, 2001).

2.   The visit to the United States was made in 1863 as a Prussian army lieutenant from the state of Württemberg. While Zeppelin is said to have observed balloon ascensions coordinated by Thaddeus C. Lowe for Union Army observation of Confederate positions, it was on a vacation in Minnesota at the end of his trip that he found his own way into the air, paying for an ascension with ex-army balloonist Jules Steiner. "While I was above St. Paul," Zeppelin would reminisce in later years, "the first idea of my Zeppelins came to me." Tom D. Crouch, *The Eagle Aloft: Two Centuries of the Balloon in America* (Washington, D.C.: Smithsonian Institution Press, 1983), 281–282.

3.   Wiley Post and Harold Gatty, *Around the World in Eight Days* (New York: Garden City Publishing, 1931), 22–23.

4.   "Post and Gatty Get Stirring Welcome," *New York Times*, July 3, 1931, 1.

5.   Wiley and Gatty, *Around the World in Eight Days*, 27.

6.   *Oklahoma City Times*, July 15, 1933. Cited in Stanley R. Mohler and Bobby H. Johnson, *Wiley Post, His Winnie Mae, and the World's First Pressure Suit*, Smithsonian Annals of Flight 8 (Washington, D.C.: Smithsonian Institution Press, 1971), 69.

7.   Major Charles L. Wilson, U.S. Air Force, "Wiley Post: First Test of High Altitude Pressure Suits in the United States," *Archives of Environmental Health* 10 (May 1965): 806.

8.   Mohler and Johnson, *Wiley Post, His Winnie Mae, and the World's First Pressure Suit*, 112.

9.   Roland Barthes, *Mythologies*, trans. Jonathan Cape (New York: Farrar, Straus and Giroux, 1972), 71–73.

10. "Post Forced Down from 24,500 Feet," *New York Times*, February 23, 1935, 1.

Layer 6 CYBORG

"SPACEMAN IS SEEN AS MAN-MACHINE"

The word "cyborg" first appeared in print on May 22, 1960. On page 31 of the *New York Times*, an article, "Spaceman Is Seen as Man-Machine,"[1] reported on an advance copy of the proceedings of a symposium to be held several days later in San Antonio, Texas: "Psychophysiological Aspects of Space Flight." The *Times* article focused on the twenty-seventh paper of twenty-eight in the proceedings (most of which focused on germane medical topics such as stress, cognition, sensation, and circadian rhythms). The paper in question, however, "Drugs, Space, and Cybernetics," proposed not only new vocabulary—the word "cyborg"—but also a new approach to the manned exploration of space.

The newspaper explained as follows:

> A possible picture of the space man of the future has emerged from a radically new approach to the problems of space medicine.
>
> According to the new view, a space man would be a human-and-then-some.
>
> He would not have to eat or breathe. Those functions and many others would be taken care of automatically by drugs and battery-powered devices, some of which would be built directly into his body.
>
> So equipped, the space man would belong to a breed of literally super-human beings that the scientists who conceived them call "cyborgs."

The word "cyborg" is so much part of our contemporary culture that to examine its singular origin seems uncanny.[2] Attendant to its remarkable birth were two doctors (only one an M.D.); the history and credit for their invention is confused not least by the similarity of their two names— Manfred Clynes and Nathan S. Kline. At the time both were based at the Rockland State Mental Hospital, perched above the Palisades a short drive from Manhattan.

NATHAN S. KLINE

In becoming a doctor at all, let alone a highly successful and public one, Kline was a rebel. His father, a self-made New York businessman, opposed his son's plans even to go to college. The irrepressible Kline, however, paid his own way through three degrees, earning a B.A. in philosophy, an M.A. in psychology, and an M.D., and undergoing a full course of Freudian psychoanalysis before qualifying as a psychiatrist. At the same time, Kline nurtured a set of important social connections, befriending among others Mary Lasker, widow of Albert Lasker and director of the prestigious award for medicine that bears their name (and which Kline became the first clinician to win twice). Displaying a fine sense of both therapy and theater, Kline first gained notoriety in medical circles by arranging a program of Shakespearian performance for terminally psychotic inmates at an asylum in Worcester, Massachusetts.

ROCKLAND STATE HOSPITAL

A photograph of the Rockland hospital from the year of Kline's arrival as research director shows a sprawling complex of buildings, covering hundreds of acres and housing tens of thousands of patients. As well as having its own power and steam plants, transportation services, and an enormous catering staff, the massive hospital complex employed a team of ten full-time glaziers to fix the windows constantly destroyed by the patients in the hospital's "back" wards, which housed those institutionalized indefinitely.

In 1952, a mere half-century separated the young Nathan Kline from the first work of Freud, as well as that of Emil Kraepelin, who in the 1890s separated the general term "insanity" into two separate disorders—manic depression and dementia praecox (today known as schizophrenia). Even in 1952 a crucial step in the history of mental illness was yet to be taken: the invention of drugs to treat the brain. To truly be considered a "disease" by modern medicine, a condition has to have a proven treatment, and for mental illness this was far from the case. The full frontal lobotomy—severing all connections to the prefrontal cortex—was losing favor in medical circles to the "partial" lobotomy, in which the cortex was damaged with a sharp needle pounded through the top of the eye socket. As with electric shock or high-therapy water treatment, however, a midcentury practitioner of the technique admitted it was done more often to aid in "management" of the patient than to produce any marked improvement in the underlying disease.[3]

In 1946, a former patient at Rockland State Hospital published a fictionalized account of her experiences, titled *The Snake Pit.* As the dust jacket explained: "Long ago men tried to shock the insane back into sanity by throwing them into a snake pit. ... Modern methods, though superficially more civilized, often rely on the same brutal shock to achieve their results."[4] The appearance of the novel—and a film version starring Olivia de Havilland—made Rockland even more of a "problem hospital"[5] in the eyes of New York State officials. A small research unit, which the young and already socially well-connected Dr. Kline was invited to lead, was proposed to "boost morale."[6]

6.1
Manfred Clynes (left) and Nathan S. Kline seated behind an electronic printout of biological measurements.

One of many programs initiated by the enterprising Dr. Kline emerged from a conversation with the Swiss pharmaceutical company Ciba. Kline approached the company initially seeking funding for equipment from his small, cash-starved laboratory. Ciba offered the funds, but with a condition—would Kline try out a new compound that might show promise for the treatment of mental disorders?

The compound, reserpine, was derived from a snake-shaped South Asian root first described to Western science by the sixteenth-century German botanist Leonhard Rauwolf—*Rauwolfia serpentina*. It had been known to traditional Vedic medicine for centuries. Noting its traditional use for hypertension, Indian chemists set about isolating its active ingredients in the late 1940s. When it was tested for the treatment of hypertension, however, patients also reported feelings of tranquility and well-being, which led to Ciba's consideration of the compound for the treatment of mental patients.

After receiving a supply of both *Rauwolfia serpentina* and reserpine from Ciba (along with the $1000 he had originally sought for a blood-gas machine), Kline set about administering the drugs to over seven hundred of his most schizophrenic and psychotic patients (the regulation of such testing by the FDA would not arrive until 1962). What happened next was historic, and Kline would often recount the story of his discovery of the drug's therapeutic effects on schizophrenics. He would explain how the drug's effects were revealed to him initially by the staff glazier, who questioned the Institute director about the sudden drop in window repairs on the ward where reserpine was being administered.[7] Miraculously, the seemingly intractable psychotic behavior of schizophrenic patients had disappeared, without the sedation, injury, or restraint of earlier treatments.

Kline published the results of his research in the *Annals of the New York Academy of Sciences*[8] in 1954, and catapulted himself into the beginnings of a revolution in psychiatric treatment.

And, despite the eventual overtaking of the natural reserpine by the synthetic (and so more profitable) chlorpromazine[9]—brand name Thorazine—it was a revolution in which Kline was to be in the vanguard.

Soon after the publication of the reserpine results, Kline continued his role as one of the most public faces of the psychiatric revolution with his testimony before Senator Lister Hill's Public Welfare Committee.[10] Kline and others were so successful in convincing the committee, and the Senate, of the importance of new research in biological psychology that the Mental Health Study Act of 1955—which resulted from their testimony—earmarked so much money for new mental health research that its grantors found it impossible to give away the entire sum in the first year of appropriations.

A "NEW LOOK" IN MENTAL HEALTH

As one of the most prominent actors in the new drama of psychopharmacology, Kline's research and laboratory expanded even as the administration of reserpine and chlorpromazine allowed the discharge of thousands of patients from Rockland and other mental institutions. Termed a "New Look" in mental health,[11] this transformation was a true revolution, especially in a field traditionally dominated by surgical and Freudian responses to mental maladies. As Dr. Kline explained later in his life, moreover, "the fact that a condition is treated with medication somehow guarantees in the public mind that it is a genuine illness";[12] the pharmaceutical treatment of insanity would lead to a broad change in the public perception of mental illness as a disequilibrium, not simply a disgrace.

As the Rockland Research Institute prospered—growing quickly to hundreds of researchers and experimental programs—success followed success. Attempting to reproduce his results with reserpine, Kline succeeded again in spectacular fashion when searching for a "mental energizer" to treat those afflicted with incapacitating depressive disorders. He tested a tuberculosis drug, iproniazid, whose mood-lifting properties had been noted by earlier researchers. When the drug was tested on Rockland patients in 1957, Kline's hunch again led to clinical results.[13] When iproniazid turned out to have severe side effects, it was mainly due to Kline's increasing public profile, and his tireless advocacy, that the class of drugs that iproniazid was a part of—monoamine oxidase inhibitors (MAOIs)—remained available to patients and researchers.

In the meantime, Rockland was becoming a professional and social center for advanced physiological psychiatric research, and Kline (who took up a position at Columbia University in addition to his Rockland work) became known to researchers in fields as diverse as engineering, pathology, and experimental biology. He placed enormous faith in the payoff of research outside normal boundaries in psychology—especially given the return on his own professional gambles. Thus, attending a lecture in 1954 by an advanced engineering student at Princeton on the potential application of new techniques in analog information processing to rocket engine control, he invited the researcher, Manfred Clynes, to Rockland.

6.2
Rockland State Hospital, central campus c. 1950.

MANFRED E. CLYNES

Compared to Kline, Manfred Clynes had grown up in relative poverty, but in a family whose emphasis on education and culture was without equal. Born to a scholarly Jewish family in Vienna and educated in Budapest, Clynes escaped with his family from Nazi-occupied Austria in 1938.[14] Months later, by the time Clynes and his family voyaged to Australia, their whole social and cultural context had been obliterated. Whether as a direct result of this earlier rupture or not, Clynes seemed never afterward to settle—physically, intellectually, or otherwise.

As his family made its new home in the immigrant haven of Melbourne, Clynes found work in an automotive parts factory before attempting to enter Melbourne University simultaneously in two disciplines—engineering and music. Astounding his tutors by completing both degrees while continuing his factory work, Clynes found yet another escape: to study at the Juilliard School on one of the United States' first Fulbright grants. Clynes was in turn dissatisfied by his 1948 Juilliard degree and sought further study in engineering and mathematics as well, arriving at Princeton in 1952. Perhaps through its resonance with his musical training (which by this time had made him a highly accomplished pianist and cellist) Clynes found himself enormously adept—and in high demand—in the new field of analog computing. Unlike digital computers, with their adamantine binary circuits, analog computers are based on the recording, transmission, and manipulation of continuous, wavering signals such as sound vibrations, energy flows, and—most importantly for Clynes and Kline—physiological data.

While digital computers eased through their extended development by IBM and other large corporations in the 1950s, analog devices were, by comparison, cheaper and easier to develop. Furthermore, interest in the manipulation of the continuous signals controlled and studied by such devices was at the forefront of scientific fashion, fueled by the emergence of a new discipline studying control and communication in natural and manmade systems—*cybernetics*.

A CYBERNETIC INTERLUDE

In 1941, the high altitude at which German planes were crossing the British coastline allowed sufficient reaction time on the part of a pilot to anti-aircraft fire such that the artilleryman's rule of thumb (fire where you expect the target to be) developed into a complex game of cat-and-mouse. Attempting to bring some of the same science to the problem already being applied to transatlantic convoys, supply lines, and other linear military operations, the Army asked MIT mathematician Norbert Wiener to address the problem. A momentous prodigy himself, Wiener had been writing on the topic of control systems since before the war.

As suggested by Wiener's invented etymology (from the Greek κυβερνητεσ, meaning helmsman), the resulting study defined *cybernetics* as the science of the artilleryman's essential variables: communication and control. Using one of the country's first electronic analog computers (Vannevar Bush's Rockefeller Differential Machine at Harvard), Wiener established a series of calculus-based equations whereby the imperfections in a controlled system (from the flaps of an airplane to a human hand reaching for an object) could be described and understood mathematically—the first step toward compensating for the particular nature of a control system when designing a prosthesis or other device to interact with the system as a whole.

While no artillery system ever emerged directly from the research (with the end of the Battle of Britain, the need for accuracy in anti-aircraft fire was decreased), the effects of the resulting theories—which could discover and describe systems of control not just in manmade mechanical systems, but also in the natural world—was enormous. The study of signals and control for their own sake may seem obvious today. But it was not until Wiener proposed such a distinction that the intelligible content of telephone wires, for example (as opposed to the surrounding static), was described as something with its own substance and behavior—Wiener's newly coined "information."[15] In this distinction between substance and content, between physical material and virtual data, Wiener defined the information revolution that was about to take place across the post-war world.

CLYNES + KLINE

Thus, when Kline (psychologist) first encountered Clynes (signals expert) in 1954, they met in a context where the study of signals, data, and "information" was at its first, frothy boil. By Clynes's recollection, they had spoken for a few minutes only when Kline invited him to join the research staff of the Rockland hospital, to study the applicability of signal and control theory to the brain. Within the year, Clynes was heading a new "Biocybernetics Laboratory" at Rockland.

While conferences on cybernetics had since the 1940s drawn from disciplines as diverse as physics, biology, and engineering,[16] the establishment of a cybernetics laboratory in a psychiatric hospital at the dawn of deinstitutionalization is particularly important to cyborg history. In light of the striking and seemingly impossible physiological transformations pioneered at Rockland to treat depression and schizophrenia, the notion of "control" imbedded in cybernetic language moved from a possibly passive prefix to an active, transformative approach to the human body. It was in this same year that Wiener himself shared concerns about the use of cybernetic ideas on man, in his book *The Human Use of Human Beings*.[17]

In a crucial sense, the deinstitutionalized patients of Rockland County hospital were cyborgs before the fact, equipped (as long as their medicines were administered) to deal with an environment—modern society—as foreign to their human maladies as the moon itself. In the context of Clynes and Kline's larger collaboration, the resulting "cyborg" construct was as natural as the systems it sought to augment.

OF CATS AND CADILLACS

After his arrival at Rockland, one of Clynes's first inventions was a device for measuring and storing the results of biological signal measurements at a previously impossible level of resolution. The machine was able to listen to a repeating signal—as in a heart rate or firing nerve—and use the similarities between repeating signals to increase the resolution of the signal as a whole. The "Computer of Average Transients" (CAT) was assembled from transistor technology, and easily replaced expensive vacuum-tube devices. Ever entrepreneurial, Kline suggested to Clynes that he establish a company to market the device. Mystified, Clynes countered that he was unsure of how to go about such a proposition; Kline returned quickly with $100,000 in capital raised from New York connections. Kline also named the company "Mnemotron" (combining the pair's first initials with a play on the Latin word for memory).

MAN REMADE
TO LIVE IN SPACE

Striding buoyantly across the low-gravity surface of the moon, there may someday be strange new men— part human, part machine—like the ones above. They will have a strange name: CYBORGS (for CYBernetic ORGanisms). Cyborgs, according to a daring new idea, will be men whose body organs and systems are automatically adjusted for life in unearthly environments by artificial organs and senses. Some of these devices will be attached, others actually implanted by surgery. With their aid cyborgs can dispense with clumsy, easy-to-puncture space suits in which earth conditions are re-created. Instead they can move about safely wearing not much more than they would at home.

The artificial senses of cyborgs will measure changes inside the body and outside in the environment. They will signal artificial glands telling them what to secrete for regulating normal body functions. Then body temperature may fall to that of a fish in ice, or the pulse may quicken like a robin's in flight, but the human organism will survive. Fantastic as the idea sounds, its originators (*see next page*) think that it is feasible and that much of the knowledge needed already exists.

ON MOON cyborgs unreel cable to explosive for seismic blast. On front cyborg's belt, tubes pump chemicals to blood to control (*from left*) blood pressure, pulse, energy, tranquillity, blood sugar, body temperature, radiation tolerance. Pumps obey sensors like radiation counter in his left thigh or blood-pressure gauge in his right thigh. Heart, in X-ray view, sends blood to the implanted converter which remakes oxygen and carbon from CO_2, taking place of lungs. On back of other cyborg are food supply (top), master fuel cell, food processor and wastes' canister.

CONTINUED

6.3

Cyborgs on the moon, *Life* magazine; an image highly influential on the popular conception of "cyborg."

Within a year and a half, the Mnemotron company of Pearl River, New York, was successfully marketing its "CAT 400" computer around the world, and was purchased by the Technical Measurements Corporation of Connecticut— all at great profit to Clynes and Kline. (A colleague recalls the two men, both short in stature, driving Kline's large new Cadillac, each barely visible behind the giant dashboard.) The device developed by Clynes would become a standard benchtop appliance for research in biology and physiology, as well as playing a role in the development of Nuclear Magnetic Resonance Imaging.[18]

THE CYBORG

When *Life* magazine followed up on the *New York Times* "cyborg" article, publishing a spread on the concept and its inventors in the July 11, 1960 issue, it pictured Clynes and Kline seated behind an enormous printout of waveforms produced by the Mnemotron (figure 6.1). The larger illustration (painted by noted sci-fi illustrator Fred Freeman) showed a pair of cyborgs on the moon, complete with silvery, hard skin, and quasi-architectural cutaways showing machine and organ enmeshed.

And yet it is the printout in front of Clynes and Kline, in which a man's actual pulse and breath rates appear next to a prediction of his pulse rate produced with the CAT, that reveals the most about the "cyborg" as conceived by the pair. This vision had much less to do with mechanical prosthetics than with the function of any manmade influence—mechanical, biological, and particularly chemical—within man's internal control systems. Just as a chemical moderation of these systems allowed a schizophrenic to function in contemporary society, so might they allow him, or any other person, to function in the (arguably) no less extraordinary context of lunar desolation.

SEXUALITY AND SPACE TRAVEL

For all of its physiological details, Kline's and Clynes's cyborg remained, at least to its authors, a fundamentally psychological proposition. While many of the metabolic variables whose manipulation is proposed are purely physical—heart rate, pressure, the ratio of blood gases—the proposed outcome from their regulation in a vacuum, and the resulting freedom from "constricting pressure suits," is couched in the soft language of psychology. The result will be "a new opening for man's spirit."

One of the most fascinating examples of the alignment and overlap between the cyborg's psychological origins and its expression as the subject of interplanetary travel is the extensive discussion—at least in Kline and Clynes's original conference paper—of sex. (By the time the paper was published widely, in the *Journal of the American Rocket Society*, the topic had been redacted.)[19] Identifying "Erotic Satisfaction" as one of the potential problems of space travel to be overcome by the cyborg framework, a twofold solution was proposed. First, the brain's "pleasure center" would be stimulated to assuage the need for sexual expression. Secondly, when sexual urges, or other strongly expressed feelings, were unwelcome, their chemical suppression would be introduced—either through automatic injection or, in a more extreme case, by a strong chemical dosage triggered from the ground.[20]

The repression of "unpredictable" nature, and the resulting extension of human possibility, was, in Clynes and Kline's terms, as well suited to astronautics as to psychiatry. Yet the two doctors were far from alone in their anxieties about an astronaut's well-being, including his sexual satisfaction. The notion that male sexual urges would have to be chemically repressed was wide-spread enough to be the subject of what passed for cartoon humor in the rocketry press—as in the August 1960 issue of *Astronautics*, where a doctor, syringe in hand, follows a heavyset, pressure-suited astronaut as he chases a wasp-waisted "astronaut-ette." The fact that female astronauts had already been summarily dismissed by congressional committee offers the tantalizing prospect that it was not even a "normal" heterosexual urge that might become uncontrollable in the freedom of orbit, but an interplanetary urge toward homosexual behavior instead.

In their paper, Kline and Clynes explicitly proposed a new kind of evolution for mankind. "In the past," they explained, "the altering of bodily functions to suit different environments was accomplished through evolution. From now on, at least in some degree, this can be achieved without alteration of heredity by suitable biochemical, physiological, and electronic manipulation of man's existing *modus vivendi*."

That the nature of the *modus vivendi* in question might resist alteration, or attempts to rigidly integrate it into a mechanical homeostasis, seems not to have been a prospect imagined by the pair. Though they suffered an intellectual and professional split only a few years later (rooted in some of Clynes's wilder assertions that human emotion could be recorded, replicated, and manipulated through "Sentic" technology),[21] each would continue to assert the potential primacy of external control systems over the body's internal functioning. In the case of Nathan Kline, this chiefly took the form of a lifelong advocacy for psychopharmacology, culminating in his bestselling 1974 manifesto on the chemical treatment of depression, *From Sad to Glad*.[22] His work on the use of computing in psychiatric treatment continued, but came to focus on the administration and statistical analysis of treatment by digital means.[23]

Manfred Clynes, on the other hand, continued his work on the electronic recording and manipulation of emotion (even after this "Sentics" research led to a split with Kline and the Rockland Hospital).[24] When asked in 1970 to submit a follow-up to the first Cyborg article printed in *Astronautics*, he submitted an extended manifesto on this research, "Sentics and Space Travel"; it was rejected.[25] Seemingly brilliant but also incomprehensible, Clynes's publications on the "biocybernetics of emotions," and their manipulation and measurement using finger and chin movements,[26] garnered memorable confusion from a *New York Times* review in 1977: "I read every word of this book but frankly I am not sure I know what Dr. Manfred Clynes is talking about."[27]

A fanciful view of chemotherapeutic control of sexual drive in manned space missions.

6.4
Cartoon from *Astronautics*, journal of the American Rocket Society, illustrating popular notions of the control of sexual drive in space.

Yet there is a spirit in Kline and Clynes's cyborg proposal that is as revealing of our current age as it is lodged in its own. The pair's 1960 article uses the word "milieu" to describe a setting to which the human body could be adapted. This word, however, is deployed in a strict reversal of its medical origins. For the doctor of the nineteenth century, operating before the success of germ theory, a specific "milieu"—such as a clean or dirty neighborhood—was an environment that introduced certain qualities to its inhabitants—diseased or healthy, tall or short, and so on. For Clynes and Kline, however, a new "milieu," a new technology-enabled age, was simply an environment to which the human mechanism, thoroughly integrated into a global technological network, as a curve on their oscilloscope display, could be adjusted, tweaked, and shaped. And while humans on the moon would never undergo such profound physiological modification, the curves of Clynes and Kline's cyborg would continue to haunt the systems architecture of the space age.

Notes

1. "Spaceman Is Seen as Man-Machine," *New York Times*, May 22, 1960, 31.

2. The term, as will become clear, was initially wildly optimistic in its outlook. Having gone through a cycle of dystopian signification (recall *Star Trek*'s 'Borg), it was in turn reinvented as an article of self-empowerment by the philosopher Donna Haraway in her influential text "A Cyborg Manifesto." See Donna Haraway, "A Cyborg Manifesto," in *Simians, Cyborgs and Women: The Reinvention of Nature* (New York: Routledge, 1991). In this chapter, I restrict myself to considering the notion as originally conceived by Clynes and Kline.

3. Nathan Kline and Kenneth E. Livingston, "Spectrum Forum: Drugs vs. Surgery for Psychosis," *Pfizer Spectrum* 5, no. 9: 263–265.

4. Mary Jane Ward, *The Snake Pit* (New York: Random House, 1946), dust jacket.

5. Brian Healy, *The Antidepressant Era* (Cambridge, MA: Harvard University Press, 1997), 104.

6. Ibid., 105.

7. The phrase "glazier" conjures the elegant repair of thin, float-glass panes. The window panes at Rockland, however, were inches thick, and protected by bars. The frequent need to repair them was a testament to the strength and determination of the hospital's psychotic patients, who were known to rip furniture from bolted brackets in their most disturbed episodes. Stuart Moss (research librarian), Terry O'Keefe (laboratory manager), and Tom Cooper (director), Analytical Psychopharmacology Division, Nathan S. Kline Institute for Psychiatric Research, interview with the author, January 29, 2004.

8. Nathan Kline, "Use of Rauwolfia Serpenthia Benth in Neuropsychiatric Conditions," *Annals of the New York Academy of Sciences* 59 (1954): 107–132.

9. The results of the clinical trials of chlorpromazine, which was developed in France and Switzerland, were published within weeks of Kline's announcement of his reserpine experiment. See Healy, *The Antidepressant Era*, 106.

10. The senator's father, an Alabama doctor, had studied in Europe with the groundbreaking public health advocate Joseph Lister, and named his son in his honor.

11. "New Look," *Medical World News*, August 4, 1967, 52.

12. Lindsey Gruson, "Nathan Kline, Developer of Antidepressants, Dies," *New York Times*, February 14, 1983, D10.

13. The second Lasker award Kline received for this research, following his first award for reserpine, would be contested by John C. Saunders, a junior researcher at Rockland, until a 1981 court order determined he should receive a third of the $10,000 award. See obituary by Wolfgang Saxon, "John Clarke Saunders, 82, Physician and Clinical Pharmacologist: A Court Case Brought Recognition for Work with a Euphoric Drug," *New York Times*, March 29, 2001, A25.

14. Manfred Clynes, telephone interview with the author, January 23, 2004.

15. Flo Conway and Jim Siegelman, *Dark Hero of the Information Age: In Search of Norbert Wiener, the Father of Cybernetics* (New York: Basic Books, 2004), 15.

16. Ibid., 128.

17. Norbert Wiener, *The Human Use of Human Beings* (New York: Anchor Books, 1954).

18. Moss, O'Keefe, Cooper, interview with the author, January 29, 2004.

19. Manfred E. Clynes and Nathan S. Kline, "Cyborgs and Space," *Astronautics*, September 1960, 26.

20. Bernhard E. Flaherty, *Psychophysiological Aspects of Space Flight* (New York: Columbia University Press, 1961), 369. Proceedings of the conference held in San Antonio, Texas, May 26–27, 1960.

21. See Manfred Clynes, "Sentics: Biocybernetics of Emotion Communication," *Annals of the New York Academy of Science* 220 (1973): 55–131.

22. Nathan Kline, *From Sad to Glad: Kline on Depression* (New York: Putnam, 1974).

23. See Nathan Kline and Eugene M. Laska, "Computers and Electronic Devices in Psychiatry," *World Congress of Psychiatry* (New York: Grune and Stratton, 1968).

24. Moss, O'Keefe, Cooper, interview with the author, January 29, 2004.

25. Clynes, interview with the author, January 23, 2004.

26. Manfred Clynes, *Sentics: The Touch of Emotions* (New York: Doubleday Anchor, 1977), 250.

27. Alix Nelson, "Nonfiction in Brief," *New York Times Book Review*, February 5, 1978, BR5.

Layer **7** FLIGHT SUIT TO SPACE SUIT

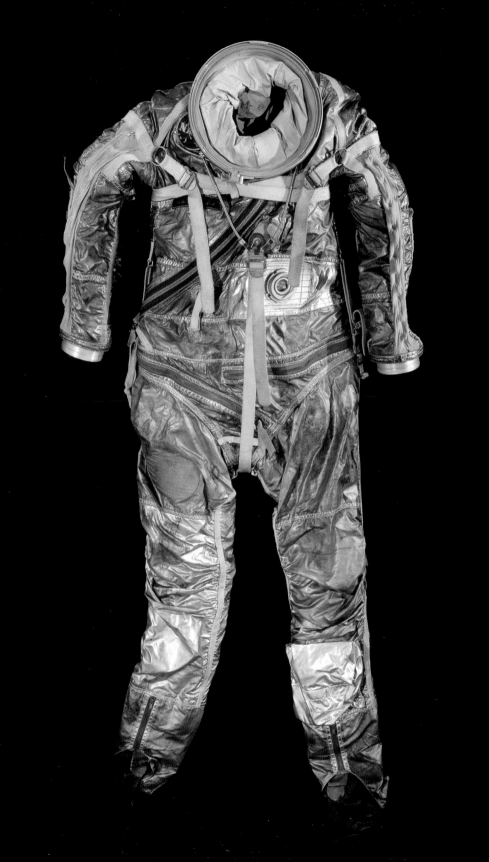

Today, Alan Shepard's Mercury pressure suit lies on a foam pallet in a refrigerated storage locker at the Smithsonian Institution's Garber Storage Facility. Originally a brilliant silver, the suit now bleeds army green and dark brown from its fabric, zippers, and openings. The greasy film that seeps through the skin of the suit dissolves the silver coating, and reveals it as a thin, sprayed-on veneer. On the suit's gloves, the corrosive glue that once bound the finish to the underlying fabric has eaten away at the aluminum such that glints of silver survive only in the deepest folds. And although the fabric tapes and laces which prevent the suit from ballooning out of shape when pressurized are a neutral white in color, its many zippers—which allow folds of the pressure bladder to hang out of the fabric restraint layer while the suit is wriggled into—are combat green.

While silvery spacesuits date from film serials such as *Flash Gordon* in the 1930s, the Mercury suit's current appearance belies this space-age image, and reveals the fact that, under their dusting of aluminum, early "space suits" were in fact identical twins of military high-altitude pressure suits, whose development in the 1940s and early 1950s stemmed from the need to fly further, faster, and higher. Survival equipment for a Cold War arms race, these suits had their origins in the unexpected limits to which unprecedented technological developments of an earlier conflict pushed soldiers, planes, and pilots.

MAN VERSUS MACHINE

In the military twilight of 1945, the German Luftwaffe started operational flights of the Messerschmitt ME-262 *Schwalbe*, or "Swallow." It did not resemble any airplane previously used in World War II, and its sleek, sinister frame held an entirely new kind of propulsion—the turbojet—that threatened to revolutionize the European conflict even as it reached its closing days. The top speed of the new fighter was three times the cruising speed of the most advanced American bombers, and outpaced the top-of-the-line Allied P-51D fighter by more than 100 miles per hour. General Carl A. Spaatz, Strategic Commander of U.S. Air Forces in Europe, described the aircraft as the "largest threat to his forces," warning in January, 1945 that the new plane could allow the German military to "establish air superiority not only over Germany but over all of Western Europe."[1]

For the Allies, the jet-powered German fighter was a grim preview of changes in aircraft propulsion and performance that would revolutionize aircraft design after the war. Yet the *Schwalbe* itself turned out not to be the threat that Spaatz anticipated. While Spaatz focused on obliterating the factories producing the planes, the pilots of the jets already in combat soon found themselves vulnerable to new tactics on the part of American pilots. In seemingly impossible dives and fast turns, the Americans whirled around and above the fast plane's trajectory, exposing it to attack from the top, bottom, and rear. Such maneuvers were not impractical for the airplanes, but they were impossible, the Germans believed, for the humans piloting them. The forces of acceleration experienced during such maneuvers, the Luftwaffe calculated, lay far beyond the range at which unconsciousness should occur.

7.1

Alan Shepard's B. F. Goodrich-produced
Mercury spacesuit.

While the Swallow had the advantage of revolutions in engines and aerodynamics, the American pilots had the advantage of a secret revolution in clothing—the antigravity, anti-G or G suit—which became standard issue for high-performance fighter pilots by the end of the war. Containing a system of inflatable bags worn strapped to the lower legs and abdomen, the suit inflated these bladders as the plane accelerated through aerial maneuvers, squeezing the pilot's pooling blood like toothpaste toward his head, and keeping him conscious and aware. Like the Apollo spacesuit decades later, the successful G-suits had their origins in underwear.

"GETTING THAT WATERMELON SILHOUETTE?"

In 1935, David Clark, the 32-year-old head knitter of the Worcester Knitting Company, left the clothing firm to start his own business with his father. In his new firm, Clark introduced an innovative knitting machine that could produce an entire elastic girdle in one piece, increasing comfort and performance and decreasing production costs. Using the patented technique, the firm initially supplied "blank" generic corsets for retail clothing firms to modify and sell as their own. In 1941, Clark advertised his first consumer product.

7.2

"Getting That Watermelon Silhouette?" Advertisement for the Straightaway Abdominal Support, marketed by the David Clark Company.

"Getting that Watermelon Silhouette?" the *New York Times* advertisement asked. The answer proposed is a "Straightaway" Abdominal Support, "the new, one-piece, patented garment that's the secret to many a smart figure." Showing a broad-shouldered, goateed man wearing only the slimming vest, the advertisement describes a "soft-knit, athletic undershirt, changing first to a flexible waist, then to a powerful web made elastic by 'Lastex' yarn."

World events were to conspire against the product's success. Within ten weeks, the United States was launched into World War II, and the engines of the consumer economy were turned to military production. When the "Straightaway's" advertising copywriter asserted: "If you've read this far, you're the one that needs the Straightaway," he would end up speaking not to self-conscious commuters, but to the aeromedical research unit of the U.S. Army Air Force.

VIOLENT MANEUVERS

As early as 1932, one of the first Navy researchers in aerial medicine proposed that a corset-like device be used to help fliers to resist the acceleration forces encountered in "dive bombing and other violent maneuvers." By 1933 the Naval Aircraft Factory was producing a special "abdominal belt" for pilots, according to clinical specifications.[2]

Eight years later, when the few Japanese dive-bombers grounded during the aerial attack on Pearl Harbor were examined, Army doctors noticed that their pilots were cocooned in elastic bandages from foot to chest, a more advanced application of the same principle. The Japanese had researched the technique with a primitive centrifuge, which whirled a man in a circle to allow controlled simulation of aerial forces.[3] By 1942 researchers at the Mayo Clinic in Rochester, Minnesota had also constructed such a centrifuge, which doubled the speed and radius of the American Air Force's own prototype.[4] Offering the services of the device and its research physicians to the government, the Mayo researchers set about a dedicated study of ways to alleviate the effects of acceleration. In collaboration with the Air Force's recently expanded Wright Field Aeromedical Laboratory,[5] research began in earnest.

By 1942 Dr. Earl Wood, the chief test subject and a researcher at the Mayo lab, was working with several corset and underwear manufacturers, including David Clark, to construct prosthetic anti-G devices. As the Mayo researchers were the first to learn, loss of consciousness resulted not simply from the pooling of blood in the extremities, but from erratic arterial blood pressure caused by loads on different parts of the body. A British attempt at a G-suit based on the pooling principle involved a man-shaped rubber bag of water, worn like overalls. The ineffective suit exhausted its wearer with overheating (especially in blistering North Africa) and subjected him to a disconcerting "floating" sensation. The devices crafted by Dr. Wood at Mayo involved air-driven pneumatic pressure on the body, which could be increased and decreased mechanically as the pilot himself moved through space and experienced the forces of acceleration.

The first anti-G suit was handmade by the Berger Brothers Company of New Haven, Connecticut (in peacetime the company manufactured posture-correcting corsets for women). The device constructed by the firm weighed over 33 pounds, but was effective in the Mayo centrifuge. The David Clark Company, originally brought into the research for their fabric expertise, quickly improved the prototype. Instead of the "rubber hot-water-bottle type bladders tied together with

7.3
Army Air Forces "pneumatic pants," or G-suit,
c. 1944. Original caption suggests: "Pants
can be inflated by lung power or by mechanical
means."

tubes" featured in the first design, Clark substituted "odd-shaped" bladders of vinyl-coated nylon that "fitted the body of the wearer."[6] Combined with a lightweight nylon to hold the apparatus together, the G-suit's weight was reduced to three pounds, and Clark was commissioned to make a production version. Pilots were initially skeptical of the contraption, but its use quickly allowed increased performance to front-line fliers, and so spread quickly. Indeed, the introduction of the G-suit in one P-51 fighter group of the Eighth Air Force resulted in a doubling of the number of "kills" per thousand flying hours from 33 to 67.[7]

A 1945 *Life* magazine article shows a version of the final device, belted around the waist and connected by garter-like straps to bands around the thighs and calves of the pilot. "The suit has five air bladders worn by the pilot like corsets over key blood centers below the waist," explains the article. "The bladders inflate automatically the instant centrifugal force begins to build up…. This increases blackout resistance to seven Gs or more so that the pilots can snap planes in tighter turns and make sharper pull-outs from dives."[8]

SWALLOWS AND COMETS

At the same time that the Luftwaffe introduced the ME-262 *Schwalbe*, it was also testing another advanced fighter, the miniature rocket-powered ME-163 *Komet.* Using the same rocket technology pioneered at Wernher von Braun's Peenemünde laboratory, the *Komet* was restricted by two natural barriers—speed and the limits of its pilot's physiology. The plane could make the closest approach yet achieved to the speed of sound—Mach 1—but along with limits on acceleration in turns, the *Komet* pilots had to be especially careful about their altitude. While the rocket engines, which did not depend on the atmosphere's oxygen for combustion, could climb far beyond the limits of the turbocharged propeller craft the fighter pursued, the plane's pilot (equipped with an oxygen mask) could survive only to an altitude of 38,000 feet. In combat maneuvers, the rapacious rocket engines would run out of fuel after eight minutes.

HANDS OF "ALARMING" SIZE

In 1944, Dr. Joseph P. Henry of the University of Southern California's (USC) school of engineering had written to the David Clark Company inviting their collaboration in expanding the G-suit to cover more of the body, and thus protect against both acceleration and altitude. While the Clark Company felt it was too caught up in the production and improvement of the G-suit for the war effort to start another venture, they sent a selection of prototypes to Dr. Henry, and suggested the sewing services of Julia Greene, who had previously been David Clark's "principal experimental seamstress," and whose husband's wartime transfer had relocated her to Los Angeles earlier that year.[9]

The "intense and dedicated" Dr. Henry proceeded to be his own experimental subject for a series of prototype suits, covering the whole body in a corset-like embrace, which by 1945 allowed him to ascend to a simulated altitude of 90,000 feet in USC's experimental pressure chamber. By this point his suit left no exposed flesh whatsoever; earlier tests above 65,000 feet in altitude caused the portions of Dr. Henry's body outside the suit to swell up and become distended in a most "alarming" manner,[10] his hands at one point growing to twice their normal size.[11]

Dr. Henry was also his own subject for a range of grueling acceleration tests, which involved the regular collection of blood samples through a needle in the forehead while undergoing a sequence of immobilizing high-acceleration maneuvers in USC's own new centrifuge. (Henry administered no anesthetic with the intrusive device, as he judged the pain response to acceleration an important part of the high-G simulations.)[12]

When the newly created U.S. Air Force used captured German technology in order to develop its own experimental rocket plane program immediately after the war—aiming to "break" the sound barrier—an operational full-body suit that would protect the pilot against both acceleration and high altitude was deemed essential. The first such suit, based on Dr. Henry's work, was manufactured by David Clark. The final improvements introduced to the suit jettisoned bladders on the arms and legs for an even tighter, custom-fitted fabric surface across the whole body. This fabric was pulled taut around the pilot's flesh by inflatable "capstans"—rubber tubes along the suit woven into its structure. The suit protected against acceleration as well as failure in cabin pressure, and was used to protect experimental X-1 rocket plane pilot Chuck Yeager in his 1948 sound barrier attempts. A production contract was awarded to David Clark for almost all of the suits' production for postwar fliers. The company produced the suits in a set of twelve sizes that could fit the mean 92 percent of pilots, and custom corset maker Berger Brothers produced one-off modifications for the outliers at either end of the scale. The suit, termed "T-1," was unveiled to the public only in 1952, when an Associated Press story trumpeted it as "resembling the popular conception of the space suit."[13]

A serendipitous consequence of David Clark Company's work on pressure suits during the war was their entrance into the postwar brassiere market in 1947 ("just another 'G' problem," joked Mr. Clark).[14] The firm's experience in developing new methods of patterning and seaming clothing across the body—necessary given the unprecedented fit and performance required of the pressure suit's skintight fabric—found its way throughout the Munsingwear family of companies (such as the "Vassarette" junior line). These developments in turn formed part of the larger explosion in lingerie technology and brands that flowered with the easing of wartime fabric restrictions. By 1948 a *Vogue* magazine spread showed "22 different bra styles for different kinds of outfits."[15]

KICKING A TIN CAN

In 1945, the crew of the Boeing B-29 who released the atomic bomb over Hiroshima wore heated sheepskin jumpsuits and oxygen masks as a backup to their pressurized cabin (the crew of the plane's predecessor, the B-17, used such suits as their entire protection against cold and high altitude). The atomic bomb dropped by the B-29 *Enola Gay* was a 9,700-pound, 10-foot-long device, and released a force equivalent to 15,000 tons of dynamite. The blast rocked the plane as if it were "kicking a tin can." Nine years later, the proposed MK-17 thermonuclear H-bomb weighed 42,000 pounds, measured 5 feet in diameter and 25 feet in length, and could be released by only one aircraft in the U.S. fleet—the Corvair B-36.[16]

On March 1, 1954, one such B-36 was deployed near Bikini Atoll in the Pacific in the first operational test of a MK-17 prototype device (see Layer 4). While the bomb itself was detonated in a "shot cab" on the ground, a B-36 aircraft was flown above the detonation to gauge the

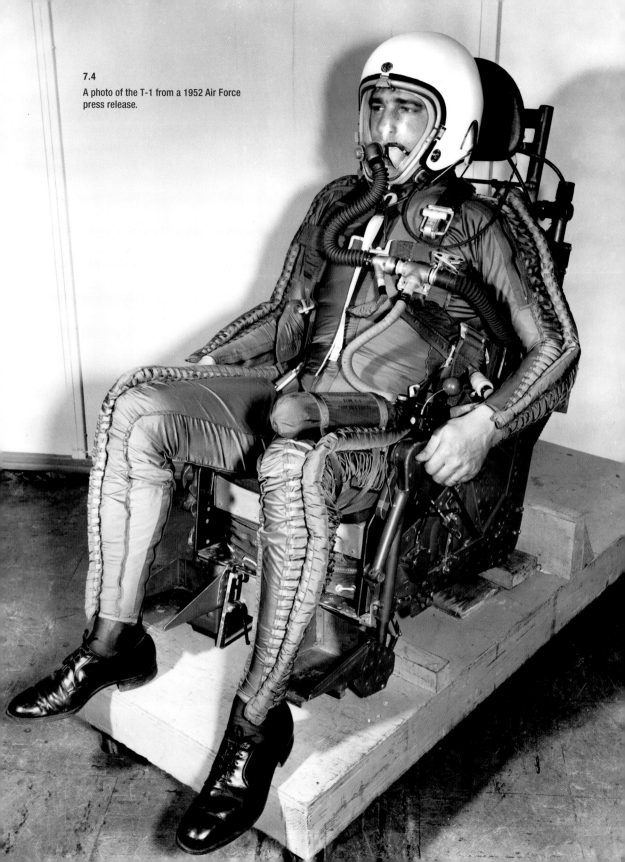

7.4
A photo of the T-1 from a 1952 Air Force press release.

potential effects of a blast on the aircraft that would drop the device in actual deployment. The test, code-named Castle Bravo, released an explosive force of more than 15 million tons of TNT. Greater than predicted, this explosion embodied one thousand times more destructive energy than the Hiroshima bomb. The B-36 barely survived. The plane was subjected to forces that doubled its designed pressure loads. Its bomb bay doors buckled inward, and dents and cracks extended throughout the airframe. The dome covering the plane's radar gear caved seven inches, and the craft's entire coat of paint was obliterated.[17] Within 60 seconds, the blast cloud itself rose higher than the B-36. Seven minutes later, the cloud rose to 130,000 feet, dwarfing the crippled airplane as its ominous underside soared tens of thousands of feet above.

This result only served to accelerate the planned introduction of the replacement for the B-36. This new jet bomber, under development since 1948, was to be the first plane designed specifically for nuclear operations.[18] As the B-52 "Stratofortress," it was first delivered for service in 1955. (The nickname, coined by a journalist, was quickly trademarked by Boeing.) As well as conveying the plane's citizenship of the thin, high atmosphere—its optimal cruising altitude was 50,000 feet—"Stratofortress" also implied a permanence and stability that politicians and manufacturers alike were eager to convey. Indeed, for the first years of its operation the B-52 was truly a permanent citizen of the stratosphere. Throughout the mid-1950s a third of the planes remained airborne at any one time, cruising the perimeter of the Soviet Union as the nation's entire nuclear deterrent. (The camouflaged missile silos of the later Cold War would await both advances in rocket technology and a reduction in the size of the huge nuclear devices themselves; see Layer 4.)

The B-52's range was almost 9,000 miles, and the concurrent development of in-air refueling made the crew's own stamina the greatest limit to its deployment. And yet almost until the last point of the plane's development, there was no technology available that could sustain the pilots for such an extended mission in the case of cabin failure. With this realization came a sudden renewed interest in pressure suit technology, as well as new orders to the Air Force's research operation to evaluate and extend the state of the art. And even as this instruction was being issued in 1954, human limits of the Cold War were being pushed further.

CURTAINS AND CEILINGS

On March 5, 1946, Winston Churchill proclaimed a new lesson in postwar geography: "From Stettin in the Baltic to Trieste in the Adriatic an *iron curtain* has descended across the Continent." While Churchill's image was a theatrical metaphor, meant to convey the "influence and … control issuing from Moscow"[19] to the newly liberated countries of Eastern Europe, the image of impenetrability it conjured was to become ever more descriptive of the physical boundaries of the Soviet Union as well.

While information had trickled out of the USSR after the war, Soviet territory was increasingly inaccessible. For a time in the late 1940s and early 1950s, brief overflights in converted bombers were attempted whenever a gap or fault in Soviet radar was detected, but these attempts at "penetration photography" ended after several U.S. planes were harassed and shot down by new Soviet jet fighters. By 1952, the vast spaces of the Soviet Union—and especially the remote areas in which nuclear and thermonuclear weapons were being developed—were effectively sealed.

It was above this literal curtain of radar and jet fighters that in 1954 the Lockheed Company proposed sending a *rara avis*, the CL-282 surveillance aircraft, which would become known as the U-2. Built to a bare half of conventional Air Force safety standards, the U-2 was a thin skin of aluminum stretched around a state-of-the-art jet engine, to which enormous sailplane wings were attached. Carrying no defenses, the plane was (at President Eisenhower's insistence) flown not by the Air Force, but by the civilian Central Intelligence Agency (CIA). Designed to carry out an entire mission while floating above 70,000 feet, the U-2 was by far the highest-altitude airplane yet developed, and forced major technical advances in every part of its operation. While it was highly classified, its development was judged a priority; creation of the plane's high-altitude fuel, for example, chemically crafted by Shell from special petroleum products otherwise used in bug spray, created a nationwide disappearance of the firm's "Flit" repellent.[20]

Operating at the limits of both man and technology, the U-2 became the first aircraft-pilot system in which a full-body pressure suit was an essential operational component. The pilot's oxygen was delivered only through his pressure suit helmet, and the pressure of the tiny cabin alone could sustain the pilot only briefly. The pressure suit was to become the pilot's entire world for an eight-hour mission, and needed to be modified to provide food, water, and other bodily necessities. The pressure suits that had been conceived as a "get me down quick" backup when first developed were becoming a sustained environment for the Cold War human body.

MISSION PROFILE: WEIGHT LOSS

The suits that emerged from the U-2's development (and also served the needs of B-52 crews) were the MC series of partial-pressure suits, again produced by David Clark. The rubber capstans of the T-1 pressure suit were joined by a massive bladder covering the pilot's entire torso, as well as inflatable gloves and tightly laced boots. Pilots were provided with water through a self-sealing straw in the pressure helmet faceplate, through which cheese- and bacon-flavored food pastes were squeezed as well. An initial attempt at providing for urination through a catheter proved painful and unsatisfactory, and a rubber, roll-on device was worn under the suit instead. No provision was made for defecation, and the pilot was given a low-bulk, high-protein diet in preparation for each mission.

The suit that resulted was described by U-2 pilots as "very uncomfortable."[21] Between the tiny cabin and tight-fitting suit, the plane's missions became eight-hour ordeals. The pilots' weight loss over the course of a mission—due to stress, the desiccating effects of breathing pure oxygen, and the tight embrace of the torso bladder—usually amounted to about 5 percent of body mass.[22] And so the U-2 marked the uncomfortable upper limit of partial-pressure suits in the developing technological race of the Cold War. The postwar flight suit's reliance on mechanically manipulating the body—pressing, molding, and holding it in shape as it sought to expand in the upper atmosphere—meant that more extreme missions and circumstances resulted directly in ever more extreme physical contortion. Such an approach had reached its literal ceiling. As a result, research into full-pressure suits—of the sort developed by Wiley Post in 1935—was given new urgency.

"NEW LOOK FOR FIGHTERS ... MEN FROM MARS"

"New Look for Fighters," trumpeted a 1948 *Los Angeles Times* photo spread. "They aren't men from Mars," the caption explained, "but they might as well be—they're U.S. officers and men dressed for the weird and terrifying forms of combat modern warfare means." Along with navy frogmen and infantrymen showing the snow gear soon to be deployed on the Korean peninsula, the largest of several images shows the "grotesque figure [of] a super-high-altitude flyer."

Unlike the corset-bound pilots of the B-52 and U-2, the figure shown is in a ballooning rubber suit, which erases the body's contours within its own distended curves. Instead of what would become known as a "partial" pressure suit, like those then manufactured by David Clark, this was a "full" pressure suit, the true heir to Post's "tire shaped like a man." As opposed to inflating bladders or capstans to put pressure on the skin of a flier, a full-pressure suit placed his entire body inside an inflated bladder, whose pressurized atmosphere (unlike the partial-pressure suit's squeezing embrace) allowed the pilot a theoretically infinite altitude ceiling.

The full-pressure suit had been the main focus of Army Air Force researchers until October 12, 1943. At this date, however, a report concluded that "not one suit was sufficiently mobile to warrant production quantities."[23] As a result, on October 29, the Army Air Force canceled its pressure suit program. It was only after a command from Air Force hierarchy the following year that the research operation restarted with an emphasis on the skintight "partial" pressure suit.

The Navy Air Services, in the meantime, continued their full-pressure suit research, collaborating with B. F. Goodrich, the producer of Wiley Post's "tire" suit. Goodrich's pressure suit team was again led by the engineer who had assisted Post with his experiments, Russell Colley. Goodrich had been involved in the Army Air Force efforts until their cancellation in 1943, and the same October 12 report concluded they were "at least a year [ahead] of all other companies on pressure suit development, having made one hundred and fifty models" ("none of which worked perfectly," the report added).[24]

The suit shown in the 1948 article was a 1943 prototype, and the result of an important insight into making the drum-tight suit bladder more mobile. In a story whose precise details may be apocryphal, Dr. Colley was said to observe a plump "tomato worm" (*Protoparce quinquemaculata*, also known as a potato worm or tobacco fly) munching the leaves of the plant he was tending in his Ohio garden. As the larva made its slow way across the plant's leaf, Colley reported his fascination with the bellows-like movement of the worm's many segments, showcased by the white stripes which danced against each other as the creature moved.

When translated to the Goodrich pressure suit, these dancing stripes became the Michelin-man segments on the arms and legs of the "flyer" shown in the 1948 photo spread. The magazine spread and accompanying illustrations of the wartime research clearly influenced popular images of pressure suits and space travel—as in Tintin cartoonist Hergé's *Destination Moon*, whose serialization began in 1950.

From the time the Air Force had suspended its own research in 1943 until the limitations of partial-pressure suits in developing Cold War operations became apparent, full-pressure suits were said to be on the "back burner" of financing and institutional involvement. The Navy retained B. F. Goodrich with a minimal contract, and the David Clark Company contributed their own work *pro bono* in the hope of future funding.

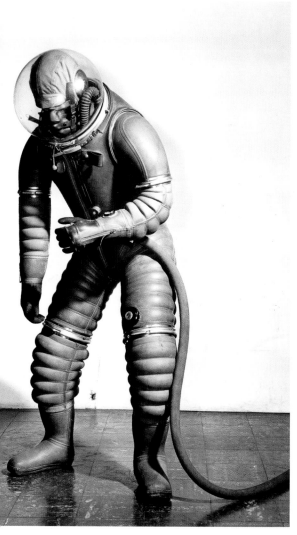

7.5

B. F. Goodrich Mark 1 "Tomato Worm" suit, showing limited mobility.

7.6

Hergé, "On a marché sur la lune."

"THE WORLD'S FIRST SPACE SUIT"

Then, the February 28, 1953 cover of *Collier's* trumpeted an exclusive look at "the world's first space suit." A reporter had been allowed to tour the Navy's full-pressure research lab in Philadelphia as part of a larger series on "Man in Space." "In such a suit," the magazine announced, "will man walk on the moon."[25]

The suit on the cover was a close relative, visually, of the 1943 suit that had appeared in the *Los Angeles Times* (and thus also the popular conception of the "space suit" which such images created). Another B. F. Goodrich design, the cover image shared with the 1943 suit a thick rubber, segmented skin, and a bubble helmet (beneath which the pilot had to wear his regular crash helmet and oxygen mask). As technicians from David Clark were chagrined to learn, the Goodrich suit had replaced a David Clark prototype on the cover—a composite fabric and rubber affair—because the "khaki-colored" Clark suit appeared far less futuristic to the magazine's editors.[26]

And yet the ugly-duckling fabric and rubber composite, not shown in the magazine, better represented the complex construction that newer suits were developing in order to solve the persistent problems of pressure suit design. The suits made by B. F. Goodrich after 1954 would also use a composite strategy of confining the unwieldy rubber bladder at the heart of the suit within a tailored fabric skin, a "constraint layer" whose pleats, folds, fabric tapes and zippers held the rubber bladder to the shape of the wearer's body while allowing restricted yet improved mobility. The David Clark suits would develop an even more multilayered strategy employing an additional restraint layer in between the suit's bladder and outer cover. With the restarting of its own full-pressure suit program in 1954, the Air Force paid $130,000 (almost one million 2010 dollars) to purchase five Navy suits, and thus study the state of the art. A subsequent report bemoaned the suits' "lack of mobility, low-comfort factor, restricted vision, poor land survival qualities [in case of ejection and separate landing by the wearer], and the extreme difficulty of getting it on and off the pilot." The suits were judged "unacceptable," the report continued, and "in September 1954, in spite of several improvements resulting from Air Force efforts, it appeared impossible for the Navy suit to meet the requirements of the swiftly developing B-52 [or the highly classified U-2.]"[27]

The account continues: "Because of the failure of the Navy suit to meet the Air Force requirements, the Air Research and Development Command on 3 February 1955 served notice that the Aero Medical Laboratory was developing a full pressure suit of independent design to fill Air Force requirements."

The Air Force first negotiated a contract with an almost entirely new entrant into the suit design business, the International Latex Company (ILC), known by its consumer brand of "Playtex" (see Layer 9). Then headquartered in the Empire State Building in Manhattan, ILC's factories were located throughout the northeast United States. A plant in Dover, Delaware produced not bras or girdles but rather the shiny, tubular containers the underwear was sold in ("The newest of the famous Playtex girdles, 'Pink Ice,'" an advertisement of the time announced, "in shimmering pink tubes!") and the metal racks which displayed the tubes in department stores. This plant had been responsible for the company's abortive 1949 bid to produce pressure suit helmets for the Air Force, and was also the source of this latest military bid from the firm.

7.7
The Navy full-pressure suit and Air Force's partial-pressure T-1 suit at a joint press event, 1953.

In the end, a major innovation by the David Clark Company in full-pressure suit design displaced ILC from the Air Force suit market. But the developmental suits that ILC constructed while under retainer from the Air Force were to have a crucial impact on the trajectory of the firm's future involvement with pressure suit research.

KNITTING THE FUTURE

By the mid-1950s, the most active test bed for suit development throughout the late 1950s was not the space program (NASA would not be founded until 1958) but the military's advanced research program, coordinated by NASA's predecessor, the National Advisory Committee for Aeronautics (NACA), in collaboration with the armed services. The first NACA test project had been the Air Force X-1, which was followed (in the tradition of rivalry between the services) by the Navy-sponsored X-2 "Skyrocket."

The Navy X-2 was built by Douglas Aircraft, and was featured on the television series *Captain Midnight* from 1954 to 1956 as the Captain's "Silver Dart" airplane. While Captain Midnight wore slacks, a bomber jacket, and a painted football helmet, Navy Pilot Scott Crossfield wore a proto- type Navy full-pressure suit when he became the first human to travel faster than twice the speed of sound in the X-2 on November 10, 1953. After piloting the X-2, Crossfield left his position as a government test pilot for a position with North American Aviation, manufacturer of the successor to X-1 and X-2, the "hypersonic" X-15 research aircraft. Like the X-1 and X-2, the X-15 was expected to have a direct effect on the subsequent design and development of military aircraft and weapons. While wearing a full- or partial-pressure suit was a matter of personal preference for the pilot of the X-2 (Crossfield's colleague Bill Bridgeman had worn a partial-pressure suit while setting the altitude record of 79,000 feet in the X-2), the X-15's expected altitude ceiling of more than 100,000 feet would bring it above 99.9 percent of the earth's atmosphere. There was no alternative to a full-pressure suit in preserving the pilot's life should cabin pressure fail.

Crossfield had been working with the Navy full-pressure suit lab, and particularly its David Clark technicians, since the X-2 project, and he was convinced the firm's work offered the best possibility for developing a truly useful full-pressure suit. It had been under his influence that the Air Force retained David Clark on a "minuscule" contract even after its main contract had gone to ILC. More importantly for the firm, Crossfield used his influence at North American, his new employer, to ensure that David Clark would be the supplier of the full-pressure suits for X-15 pilots. The suits were judged such an essential part of the airplane's operation that they would be pro- vided by North American as an integral system of the final X-15, with David Clark a subcontractor.

In 1956, soon after the first mock-up of the X-15 was completed, Crossfield visited David Clark in his Worcester, Massachusetts factory. Clark showed him a sample of the material he had knitted by hand "on an airplane when I was going up to Alaska last month to see my daughter." The material Clark had knitted was not wool or cotton but, rather, high-performance nylon. Despite its humble origins on knitting needles, the fabric was given the high-tech name "link-net." As with a loose-knit sweater, the fabric had the essential property of being able to change its geometry without alteration to the volume it contained. As soon as it was substituted for the heavy convolu- tions and structural rubber bladders of the firm's previous suits, the weight of the entire pressure suit dropped from 110 to 25 pounds, and its mobility while pressurized "vastly" increased. As Crossfield would explain: "the link-net material proved to be the great 'break-through' in the full-pressure suit game"—it allowed an uninflated suit to be worn for long periods of time, and furthermore allowed a previously unprecedented amount of mobility in the suit while it was filled with air.[28] The David Clark link-net-equipped suit displaced the ILC suit in the Air Force's affections, and it has since become standard issue for Air Force high-altitude pilots (remaining the basis for the orange suits worn by post-*Challenger* Space Shuttle astronauts for safety during launch and landing).

In the story of the suit's success, however, the most visible role was played by a very different kind of fabric, a "kind of silver lamé" encountered by Crossfield—as he tells the story—on the very factory visit during which Clark displayed the link-net prototype.

Seeing the silver fabric on a worktable, Crossfield asked Clark about it. Crossfield's 1960 biography records Clark's response:

> "That's a piece of nylon with a vacuum-blasted aluminum coating. Just something one of the boys was trying out."
>
> "Pretty glamorous looking."
>
> Then a light went on in the back of my mind. "Say, Dave, why don't you make the outer cover of the pressure suit out of this material, in place of that awful-looking khaki coverall?"
>
> "Whatever for?"
>
> "You remember that time down at [the Navy's Philadelphia suit-research lab] when they took that picture for the magazine cover [the 1953 *Collier's* feature]? We don't want to make that mistake again. A coverall of this material would look real good, like a space suit should—photogenic. To justify it technically we can tell them this silver material is specifically designed to radiate heat or something."
>
> "A marvelous idea, Scotty ... I'll make the boots and gloves out of black material for contrast."
>
> "'Great touch!'" Crossfield averred.[29]

"SPACE SUIT FOR MOON JOURNEY DEMONSTRATED"

And so, when the X-15 suit made its first public appearance, it was covered in silver fabric. A November 30, 1957 *Los Angeles Times* story covering the unveiling trumpeted a "Space Suit for Moon Journey Demonstrated," and accompanied its account of the "heat-resistant" suit with a full-length photo. By January of 1958, Crossfield appeared in the suit—complete with silver overboots—on the cover of *Life* magazine, testing the "heat-resistant" suit in an oven-like "test chamber."

While the David Clark team was covering its suit in silver, the Goodrich team—which was busy improving the Navy high-altitude suit—produced a suit in late 1957 covered in its own "heat-resistant" gold. The gold fabric was dropped because of cost, but when Goodrich was contracted to provide the first Mercury "space suits" in 1959, they, like the Clark suits, were coated in aluminum powder.[30]

7.8
The Air Force photograph of a staged "heat test" that appeared on the cover of *Life* in January 1958.

7.9

Gordon Cooper just after the seal is broken
on his Mercury capsule aboard the carrier
USS *Kearsarge*, 130 kilometers southwest
of Midway Island in the Pacific.

As revealed in their old age, under their silver skin the Mercury suits were almost identical to the Mark IV naval pressure suits upon which they were based. Like the military pressure suits, they were designed not as a full-time environment for the flier, but rather as a backup in case of airplane cabin pressure loss. While even the pilot of the X-15 airplane in his David Clark suit had to fly with his helmet visor down to breathe—his cabin was only lightly pressured with pure nitrogen—the Mercury astronauts wore visors down and suits pressurized only during takeoff. And, while the silver covering was the most substantial visual difference between the Navy Mark IV and Mercury pressure suits manufactured by Goodrich, the nylon underneath the silver coating remained dark, camouflage green, as did the zippered openings essential to expanding the tight fabric constraint layer enough to be able to put it on.

Later Mercury suits, including Gordon Cooper's *Faith 7* suit, were modified on a one-off basis to provide greater comfort and wearability. For Cooper—who had to undergo a 34-hour mission designed to test his endurance—the personalized suit included white fabric panels designed to make the garment more comfortable to sit in, as well as the replacement of the stiff combat boots worn on earlier missions. Nevertheless, the suit did not take well to extended wear, and difficulties in removing perspiration buildup produced significant discomfort. According to the official mission report, Cooper emerged from his suit with the "white, wrinkled appearance characteristic of prolonged submersion in water."[31] "'Dishpan Hands,'" *The New York Times* reported, were becoming "a Peril in Space."[32]

When NASA requested proposals for the Gemini series of pressure suits—in which the first American "spacewalk" would be made—both Goodrich and Clark submitted silver prototypes, which were subjected to a rigorous program of testing, especially for mobility and comfort while inflated. Not only would the American astronaut making the first spacewalk be protected only by the pressurized suit while exiting the Gemini spacecraft, the Gemini capsule's design—dubbed a "2-seater convertible" by the astronauts—meant that even the astronaut remaining "seated" in the evacuated capsule would have to operate with a pressurized suit as his sole protection against interplanetary space.

In the competition for Gemini suits, David Clark suit proved the victor. Its link-net-bound bladder appeared more mobile than the Goodrich suit. And yet, when in June of 1965 the first launch-to-landing broadcast of a NASA mission showed Ed White becoming the first American to "walk" in space[33] (the Russian Alexei Leonov had accomplished the same feat two months earlier), the suit he was wearing was not aluminized silver, but the same white high-temperature nylon as the link-net underneath it. The high-temperature nylon provided a superior surface, and did not run the risk of the astronaut dazzling himself with his clothing while facing unfiltered sunlight. The spacewalk was widely observed, attracting more television viewers than any other single program since coverage of John F. Kennedy's assassination.

As it transpired, the first and only "silver" spacesuit worn in the vacuum of space was donned by White's successor in Gemini 9, Gene Cernan. The suit's multilayered torso was white, but the trousers of the cover layer were coated with stainless steel "Chromel-R" fabric. It was thought that the legs of Cernan's suit would be exposed to the hot rocket exhaust of the first backpack-style Astronaut Maneuvering Unit, to be tested on the flight—in the end, however, the AMU spacewalk

7.10

Astronaut Ed White, June 1965—the first
American "spacewalk" was planned as a part
of the Gemini 4 mission.

7.11
Gene Cernan after his spacewalk, photographed
by crewmate Tom Stafford. The upper surface
of his suit was white, the legs a stainless
steel fabric.

was scrubbed. Unlike the "silver lamé" of the early Clark and Goodrich suits, the fabric was enormously expensive—hundreds of dollars per square inch—and difficult to work with. Its use in later suits was restricted to areas where its alloyed strength and resistance to puncture were judged essential (as in the lunar overshoes and palms of Apollo suits). Even with its vaunted link-net layer, however, Cernan's suit proved so exhausting to maneuver during the spacewalk that the astronaut lost 10.5 pounds during the course of the mission. As with Gordon Cooper's Mercury suit, the pressure bladder filled with perspiration, which, with nowhere to evaporate, gave Cernan a night of chills in orbit before splashdown the next day. After landing, suit technicians removed a pound of sweat from each suit leg alone.[34]

From today's perspective, then, the silver spacesuit of the Mercury era seems to relate more to Flash Gordon's space uniform, or even Captain Midnight's pilot costume—images which fulfilled and reinforced the American audience's preconceptions about a "space age." Instead of the "Skelly Oil"or "Ovaltine Drinks" behind Captain Midnight's promotions (which included "Junior Flight Patrol" membership and "Secret Decoder Badges"), the silver surface of the early full-pressure suits sought to clothe space travel itself, and attempted to ensure the success of the innovations in wearability, mobility, and functionality that lay beneath the suits' deepening layers.

The media presentation of partial-pressure suits as "space suits" and the presentation of the X-15 suit as "ready for the moon" together highlight the way in which early pressure suit development was part of a wider cultural understanding of the heroic "space suit." These early, silver prototypes dominated public perception of space-age attire throughout the 1940s and 1950s. It was only really with the white-suited spacewalks of the Gemini era, and the first televised images of an American silhouetted against stars and the planet, that a paradigmatic shift in perception took place, and the aesthetics and presentation of manned, government-funded spaceflight eclipsed a prior narrative of the "space age." From this moment, the mythic quality of spaceflight ceases to follow science fiction, and instead is extended, incredibly, by science fact.

Notes

1. Ferenc A. Vajda and Peter G. Dancey, *German Aircraft Industry and Production 1933–1945* (Warrendale, PA: SAE, 1998), 96.

2. Eugene M. Emme, *Aeronautics and Astronautics: An American Chronology of Science and Technology in the Exploration of Space, 1915–1960* (Washington, D.C.: National Aeronautics and Space Administration, 1961), 32.

3. Norman Berlinger, "The War against Gravity," *American Heritage of Invention and Technology* (Spring 2004).

4. William J. White, *A History of the Centrifuge in Aerospace Medicine* (Santa Monica: Douglas Aircraft Company Missile and Space Systems Division, Advance Space Technology, Biotechnology Branch, 1962).

5. From a nadir in the 1930s caused by depression-era finances and the politics of neutrality, the military in general, and the Army Air Corps (becoming the Army Air Forces in 1941) in particular, found itself showered with R&D resources as of mid-1940. In a hearing that followed the fall of France in June 1940 (which had been made possible by Germany's advances in aerial technology),

General "Hap" Arnold, commander of the Army Air Forces, was told by Senator Henry Cabot Lodge, Jr. in a hearing on finances: "All you have to do is ask for it." The Army Air Force took the opportunity to greatly expand its research unit into a full-scale laboratory at Wright Field in Dayton, Ohio. With the onset of war, the new research focus extended throughout private industry and nonprofit institutions. Quote from Air Training Command Pamphlet 190-1, "A History of the United States Air Force" (Randolph Air Force Base, Air Training Command, 1961).

6. Lloyd Mallan, *Suiting Up for Space* (New York: John Day, 1971), 100.

7. Berlinger, "The War against Gravity," 21.

8. "Anti-Blackout Suit," *Life*, January 29, 1945, 59.

9. David M. Clark, "On Partial Pressure Suits," internal memorandum, David Clark Company, June 22, 1961, 5.

10. Ibid.

11. Mallan, *Suiting Up for Space*, 77.

12. Ibid., 91.

13. The Associated Press, "Space Suit Developed for American Air Force," October 5, 1952. See for example *Los Angeles Times*, October 5, 1952, 20.

14. Mallan, *Suiting Up for Space*, 90.

15. "Fresh Problems: 22 Answers," *Vogue*, June 1948, 145–146. Referenced in Jane Farrell-Beck and Colleen Gau, *Uplift: The Bra in America* (Philadelphia: University of Pennsylvania Press, 2002), 114.

16. The largest land-based propeller plane ever built, the Corvair had emerged from wartime research in parallel with the B-29, but was not delivered to the Air Force until just after August 1945. With its backward-pointing propellers and vestigial machine-gun turrets, the B-36 appears a strange hybrid between our image of World War II aircraft and the sleek lines of midcentury. A test bed for experimentation throughout its life, the B-36's variations included the small XF-85 "parasite fighter," which ejected from and reentered the bomb's fuselage, and the 1954 installation of a nuclear reactor on board (see Layer 19). Dennis R. Jenkins, *Convair B-36 "Peacemaker,"* Warbird Tech Series 24 (North Branch, MN: Specialty Press, 1999).

17. Jenkins, *Convair B-36 "Peacemaker,"* 79.

18. A smaller jet bomber, the B-47 had been in service since 1948, but it was too small for the massive, experimental H-bombs. See Jan Tegler, *B-47 Stratojet: Boeing's Brilliant Bomber* (New York: McGraw-Hill, 2000).

19. Winston Churchill, *The Sinews of Peace: Post War Speeches* (New York: Houghton Mifflin, 1949), 24–37.

20. Gregory W. Pedlow and Donald E. Welzenbach, "The CIA and the U-2 Program, 1954–1974," declassified CIA report, Center for the Study of Intelligence (1998), 62.

21. Ibid.

22. Ibid.

23. "Case History of Pressure Suits," *Army Air Force Official Report*, October 12, 1943. Quoted in Mallan, *Suiting Up for Space*, 110.

24. Army Air Force, from Mallan, *Suiting Up for Space*, 113.

25. Cornelius Ryan, ed., "Man's Survival in Space," *Collier's*, February 28, 1953, 40.

26. Scott A. Crossfield with Clay Blair, *Always Another Dawn: The Story of a Rocket Test Pilot* (New York: World Publishing, 1960), 241.

27. Mallan, *Suiting Up for Space*, 130.

28. While more mobile than previous suits, it should be clear that the link-net suit, later used in Gemini, was not "mobile" by any other definition. One former Gemini astronaut compared the load of the inflated suit on any individual joint of the body to holding a ten-pound dumbbell without rest. Furthermore, the work required of motion in the pressure suit produced heat, and sweat, which the suit was poorly equipped to dispose of. From interview by author with Tom Stafford, Smithsonian Garber Center, Suitland, Maryland, February 6, 2006.

29. See Crossfield, *Always Another Dawn*, 255.

30. A NASA press release explained, "the astronaut suit will be silver coated," the coating itself described as a combined "heat buffer and radiation shield," from Mallan, *Suiting Up for Space*, 188. Writing in *Life* magazine, astronaut Wally Schirra speaks of the suit's silver skin and "silvery boots," as well as its protection against heat and radiation. From Walter Schirra, "A Suit Tailor Made for Space," *Life*, August 1, 1960, 36–42.

31. "Postlaunch Memorandum Report for Mercury-Atlas No. 9 (MA-9): Part I, Mission Analysis," MSC (June 24, 1963).

32. Richard Witkin, "'Dishpan Hands' a Peril in Space," *New York Times*, May 3, 1964, 61.

33. Some NASA officials cautioned against White's Gemini spacewalk, arguing for a more tentative opening of the craft doors: "we shouldn't be putting guys in a vacuum," one is said to have argued, "with nothing between them but the little old lady from Worcester and her glue pot." Lyndon Johnson himself is supposed to have snarled in response: "If the guy can stick his head out, he can take a walk. I want to see an American EVA." See Lindsay Hamish, *Tracking Apollo to the Moon* (New York: Springer, 2001), 101.

34. Interview by author with Tom Stafford, Smithsonian Garber Center, Suitland, Maryland, February 6, 2006.

Layer 8 MAN IN SPACE

8.1
An official photograph of Gagarin in his SK-1
pressure suit on the way to launch.

According to a statement released by the Soviet Union's TASS news service on April 12, 1961:

> After the successful completion of the planned investigations and of the flight programme, the Soviet spaceship Vostok made a safe landing at the predetermined place on April 12, 1961 at 10:55 Moscow time. The pilot cosmonaut Major Gagarin made the following statement: "I beg to report to the Party, to the government, and to Nikita Sergeyevich Khrushchev personally, that the landing was normal, that I am feeling well and have sustained no injuries or disturbances." The accomplishment of a manned spaceflight opens grandiose perspectives for man's conquest of the cosmos.[1]

As well as being the first human in space, Yuri Gagarin was the first spaceman to land on earth contained only in a spacesuit.

In mid-1958, the earliest days of planning for Soviet manned spaceflight, Chief Designer Sergei Korolev considered a variety of strategies for soft-landing the Soyuz capsule on the Soviet landmass (U.S. naval dominion over the Atlantic and Pacific precluded an American-style splash-down). The parachutes needed to slow a heavy capsule to survivable speed would be too large to be lifted by the Vostok missile's thrust. A system of rocket brakes was considered, but also judged too heavy. In a characteristic move, Soviet engineers instead opted for a radical alternative.[2]

Instead of landing on the ground in his spaceship, the Vostok cosmonaut would eject at high altitude. A pressurized oxygen supply integrated into an ejection seat would inflate the cosmonaut's pressure suit and sustain life while he plummeted earthward. Nearer the ground, the cosmonaut would open a parachute and land like a paratrooper. Since access to the Soyuz capsule after the ejection was uncertain, special care was taken to ensure that the cosmonaut's suit could sustain him not only in space, but also in the extreme climates encountered across the million-acre landing zone.

The particulars of this process, and the fact of Gagarin's separate landing, were not revealed until 1978.[3] Apart from the Soviet state's reflexive secrecy, there were both pragmatic and poetic reasons for the cover-up. Pragmatically, the Fédération Aéronautique Internationale (the Lausanne-based authority which had governed aerial record-keeping since 1906) disallowed such a procedure in its record qualifications. In order to qualify as the first orbital flight of the earth, the "spaceship pilot" needed to land inside his craft.

More poetically—and long after Gagarin's primacy as the first human in space was assured—Soviet publications continued to allege the philosophical unity of pilot and vessel. A 1971 caption to the Vostok capsule attests: "The parachuted capsule in which Yuri Gagarin returned to the earth from the first space flight at 10:55 Moscow time on 12 April 1961."[4] A spacesuit—or, as it was termed by Soviet publicists, a "space uniform"—is an extension and covering of a single man, without the many implications of extrapersonal significance, collective achievement, and seemingly infinite reproducibility of a vehicle's flight. Casting the capsule as primary established in particular the supremacy of its authorship by the Soviet state. (Sergei Korolev, the gifted engineer and manager of the Soviet space effort, was credited only as an anonymous "Chief Designer" until his death in 1966.)

8.2
A Vostok capsule, shown after its separate landing,
in an undated Soviet photograph.

SOVIET SEWING

And what of Gagarin's suit itself? An examination of this singular artifact provides a useful window into the radically different strategies for human spaceflight proposed by Soviet engineers.[5] The designs succeeded on the global stage by eschewing the optimization and controlled competition of NASA contracts for simplicity of function and robustness of form.

Instead of a molded neoprene pressure barrier, for example, the Soviet suit was pattern-sewn from latex-soaked canvas. As a result (and by design), the suit leaked constantly. This loss was managed by constantly filling the suit from an air supply built into the cosmonaut's ejection seat. Like U.S. Mercury clothing, the Vostok suit was designed as a backup during flight, primarily needed to provide several minutes of oxygen during high altitude. Thus, the constant loss of air was deemed a more effective solution than the complex assembly of a less porous pressure bladder.

The Soviet cosmonaut entered his suit, and relieved himself, through extruded tubes of the same rubbery fabric, which extended from the suit's abdomen and were tied and clipped, like bread bags, to provide an airtight seal.[6] By contrast, the complex zippers sealing U.S. suits were not considered fully reliable until 1965. The Soviet entry system has proven so effective that it still adorns the Russian KV-2 suit used by Russian astronauts on their way to the International Space Station.

What appears to be a separate helmet on Gagarin's suit was in fact a protective shell around the inner fabric bladder, which attached directly to a transparent visor to avoid the potential complication of a pressurized helmet joint. The most iconic element of Gagarin's costume, the stylized script spelling "CCCP" (Cyrillic initials for the USSR), was painted directly onto the shell after Gagarin had donned his suit by suit engineer Victor Davidyantz; at the last minute, Davidyantz realized that Gagarin, if he lost consciousness during his parachute landing, would have no outward indication of his nationality, and so might be confused with a downed American spy-plane pilot.[7]

A TIGHT FIT

The first Soviet cosmonauts to land in their spaceship (as opposed to a suit alone) did not wear spacesuits at all. On October 12, 1964, the Soviet Union launched the first three-man space capsule, the Voshkod 1—a dramatic victory in the space race, especially given that the two-man American Gemini capsule would not be launched into space for another several months.

Years later, it was revealed that the Voshkod capsule was identical to the one-man Vostok capsule that had contained the first six Russian cosmonauts, but had been modified to triple its human load. The modification of a standard capsule was not unusual in the Soviet context (all Soviet satellites and spaceships used a common chassis), but the Vostok/Voshkod modification was an explicitly political move—and potentially counterproductive in the development of Soviet space technology as a whole.[8]

When three cosmonauts were inserted into the capsule, there was no room for scientific or flight instruments. As with previous Soviet flights, the capsule was controlled from the ground; now, however, there was no possibility of controlling the craft should radio contact be lost. (Gagarin's capsule had contained a sealed envelope with codes to unlock shipborne controls.) The crew also went without pressure suits, and without anything but the most elementary acceleration couches. Without room for the ejection mechanism that previous Vostok cosmonauts used, Soviet engineers installed a combination of parachute and rocket braking systems, tested successfully a week before Voshkod 1's launch.[9] Instead of pressure suits, the crew wore woolen suits and slippers.[10]

The only other Voshkod flight was equally a direct response to the United States, in this case to the two-man Gemini's projected "spacewalk"—or extravehicular activity (EVA). The second Vostok capsule was again modified, but this time to include a crew of two cosmonauts and an expandable fabric airlock that would allow a suited cosmonaut to enter and leave the ship while his compatriot remained in a breathable atmosphere. The suit worn by Alexei Leonov was a modification of Gagarin's SK-1 suit, with an additional bladder to ensure reliability (similar to the dual-bladder system Hamilton Standard proposed for Apollo's lunar suit). The additional bladder, however, proved enormously restrictive. Combined with the tendency of the suit to expand in length under weightless conditions, it made maneuvering in orbit enormously challenging. Despite three years of intensive training—including bicycling more than one thousand kilometers to improve his endurance—Leonov found the experience of his 15-minute spacewalk an ordeal. As he later recalled:

> Near the end of my walk, I realized that my feet had pulled out of my shoes and my hands had pulled away from my gloves. My entire suit stretched so much that my hands and feet appeared to shrink. I was unable to control them. It was as if I had never tried the suit on even once.[11]

The problems with Leonov's suit interfered with his ability to enter the fabric airlock. Abandoning procedure, he partially deflated his suit and entered the airlock headfirst. Once inside, he "violated instructions," opened his helmet and turned a somersault inside the two-foot-wide space to thread himself back into the narrow confines of the Vostok/Voshkod capsule. "I couldn't see anything," Leonov remembered. "I was drenched in sweat. I was exhausted."[12]

8.3

The Voshkod 1 astronauts Boris Volynov, Vasili Lazarev, and Georgi Katys, photographed shortly after landing and wearing the same clothes they wore inside the Voshkod capsule.

8.4

A series of images released by the Soviet Union of Leonov's spacewalk, 1964.

The achievement of a Soviet spacewalk, and images of Leonov, resounded worldwide. Whatever the astonishing pragmatics hidden by the Soviet shroud of secrecy, the image of the Soviet space program in early 1965 was that of an organization moving from strength to strength. And on the media battlefield of the midcentury Cold War, images ranging from Gagarin's spacesuit-framed smile to Leonov's spacewalk were devastatingly effective. It was from a position of perceived weakness, then, that planning efforts for Apollo, including the all-important lunar spacesuit, developed throughout most of the 1960s.

Notes

1. Peter L. Smolders, *Soviets in Space* (London: Lutterworth Press, 1973), 113.

2. Asif A. Siddiqi, *Challenge to Apollo: The Soviet Union and the Space Race, 1945–1974*, NASA History Series (Washington, D.C.: National Aeronautics and Space Administration, NASA History Division, Office of Policy and Plans, 2000), 192.

3. Tim Furniss, *Space Flight: The Records* (London: Guinness Superlatives, 1985), 7, 35.

4. Evgeny Ivanovich Riabchikov and Colonel General Nikolai P. Kamanin, eds., *Russians in Space* (Moscow: Novosti Press Agency; Garden City: Doubleday, 1971), 113.

5. For a complete history, see I. P. Abramov and Å. Ingemar Skoog, *Russian Spacesuits* (London: Springer, 2003). Abramov is a long-time engineer for Zvezda, manufacturer of Soviet and Russian spacesuits.

6. Interview and examination of Russian Sokol KV-2 suit worn by Dennis Tito, Cathy Lewis, and Amanda Young, Garber Conservation Facility, Smithsonian Institution, National Air and Space Museum, April 6, 2006.

7. It had been less than a year since the widely publicized capture of U.S. pilot Francis Gary Powers and, as would be customary, no announcement of the Soviet flight was made before launch. Abramov and Skoog, *Russian Spacesuits*, 48.

8. Siddiqi, *Challenge to Apollo*, 384–385.

9. Ibid., 422.

10. While the flight of Voshkod was anything but routine, later Soviet orbital flights were judged to be sufficiently commonplace that they too dispensed with backup pressure suits. This practice ended with the tragic death of cosmonauts Giorgi Dovbrovolksi, Viktor Patsayev, and Vladislav Volkov, whose capsule depressurized shortly after departing from the inaugural visit to the Salyut space station on June 29, 1971. Subsequent Soyuz capsules were modified to carry two cosmonauts in backup pressure suits. Three-cosmonaut flight resumed in 1980 with a Soyuz redesign that allowed three suited crew members.

11. Thomas O'Toole, "The Man Who Didn't Walk on the Moon," *New York Times Magazine*, July 17, 1994, 26–29.

12. Ibid.

Layer **9**   BRAS AND THE BATTLEFIELD

The patented, **five-ounce**, light-as-air

Playtex*
Living
Girdle

MADE OF SMOOTH LIQUID LATEX
"A NATURE-SKIN† THAT MOLDS YOU IN"

Not an outdated rubber garment, but a revolutionary method of curve control.

Gives with every motion of your body . . . actually lives and breathes with you!

First of its kind! Utterly unlike any girdle you've ever worn . . . feels like your own skin, weighs less than five ounces . . . light as a breeze! Not a corset . . . not an outdated rubber garment, but a revolutionary, modern method of curve-control as natural as your own lines, slimmed down.

Amazingly different! There are no seams, no stitches, no bones, no metal . . . here's a one-piece sheath of smooth liquid latex that actually lives and breathes with you! Not just a two-way stretch, but an ALL-way stretch. Gives with every motion of your body.

Becomes a part of you . . . so resilient, it flexes with your every muscle . . . controls, but does not constrict or bind!

The ideal all-occasion girdle that makes you inches slimmer in everything from an evening dress to a bathing suit! As comfortable for golf or driving as for hours of sitting at an office desk. Work or play, winter or summer, the Playtex *Living* Girdle never tires you! Always fresh . . . rinse in suds, pat with a towel and it's ready for instant use! Even removes the necessity of a sanitary belt . . . the smooth seamless panty-crotch holds the pad securely and protectively.

Delicately flower scented, in three colors . . . blossom pink, gardenia white, forget-me-not blue. Extra small, small, medium, and large. Your department store has these sizes. Or use the handy coupon below. Individually packaged in **SLIM** silver tubes. **$2.00**

Slims you for everything from evening gown to bathing suit.

Slims the hips. Preserves that casual look so smart in today's clothes.

Slims you for everything from evening gown to bathing suit.

Not a two-way stretch but an ALL-way stretch that firms you wherever it touches.

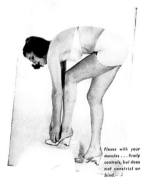

Flexes with your muscles . . . firmly controls, but does not constrict or bind.

9.1
Playtex "Living Girdle" advertisement,
Life, April 1, 1940.

FLEXIBLE FOUNDATIONS

"Without foundations," we saw Dior declare, "there can be no fashion."[1] Such foundations, of course, were the body-altering garments of which Playtex was the largest postwar purveyor. "Playtex" itself was born as a consumer brand from the foundations of its corporate parent, the International Latex Corporation (ILC). ILC's even more famous product—the A7L spacesuit—can be said in many ways to have Playtex's line of girdles, bras, and "shapewear" as its own foundation.

The ILC A7L spacesuit would not have been possible at all if not for ILC's patents and expertise in latex molding and manufacturing. These innovations had their true roots in the vanities and prejudices of postwar fashion, in particular the distortion and deformation of women's bodies in the tight rubber girdles expertly marketed by ILC's founder, Abram Nathaniel Spanel.

LIVES AND BREATHES WITH YOU

"A New Discovery in Figure Control!," announced a full-page advertisement on page six of *Life* magazine, April 1, 1940 for the "Playtex *Living* [sic] Girdle … A Nature-Skin that Molds you in!"

The rubber device professed to be "The first of its kind! Utterly unlike any girdle you've ever worn … feels like your own skin … no seams, no stitches, no bones, no metal … a one-piece sheath of smooth liquid latex that actually lives and breathes with you!" More prominent than any description are six photos of a woman wearing the girdle. In one, it is hidden underneath a tight black dress. In the others, however, it is open to view; two photos show the model naked save for girdle and shoes. Thus revealed—quite literally—was the revolution in women's underwear that arrived with the first half of the twentieth century.

As the 1900s dawned, most women of fashion still wore tight, whalebone corsets that rigidly restricted their hips and stomachs, and supported breasts only by flattening them together and pushing upward (medical experiments in 1887 recorded pressures up to 85 psi against the skin inside such corsets, the same as a modern fire hydrant).[2] Only with the 1910s did changing aesthetics, as well as a greater concern for women's health, mean a separation between garments to hold and shape the breasts, and smaller corsets and girdles around the waist and hips. So arrived the "brassiere"—first sold at its own Macy's counter in 1911—and a profusion of additional devices for the stomach, legs, hips, and chest. This separation of the corset's role into multiple garments—termed brassieres, but also "Kestoses" and "corselettes"—produced new terms for the corset's replacements, but such garments remained the "foundations" of a woman's appearance.[3] Each subsequent change in the desired female facade brought with it changes in foundation garments—and, with the growth of middle-class fashion, a widening market for underwear. In the 1920s, the industry shaped the flat line of "flapper" fashion with garments derived from medical technology. With the return to curves in the 1930s came a refinement of "breast girdle" technology, the 1933 definition of "common sizes of breasts" A, B, C, and D, and the development of waist-hugging "hip girdles."

In the 1910s and 1920s, a profusion of new magazines such as *Ladies Home Journal*, *McCall's*, and the appropriately named *Delineator* provided instruction on how to wear foundations, and which new line to aspire to. When photo magazines like *Life* and *Look* appeared in the mid-1930s, their pages—including advertisements—were aimed at both men and women. While the men's advertisements were photographed from life, emphasizing practicality (as in the depiction of a locker-room debate on "briefs over boxers"), women's bodies were depicted in a much more stylized fashion, airbrushed or cropped. *Life* articles advised women on the correct geometry to aspire to when purchasing a "bandeau" (a "brassiere," the magazine explained, extends to cover the diaphragm), and reported that sales of the garments had risen to the untold heights of 500,000 per year.

In both articles and advertisements, the link between changing styles and the "foundations" that supported them was explicit. "New Styles scared me pink till I tried the 'Press and Lift' test!" remarked one woman to another in a "Spirella" Corset promotion of 1940. As with many contemporary advertisements, women are encouraged to send for a pamphlet explaining the product in further detail; these carried titles such as "The New Art of Figure Grooming," or in the case of a more explicit competitor, "Not for the Very Thin."

The April 1, 1940 "Playtex *Living* Girdle" advertisement was a stark contrast. Instead of showing line-drawn women, or carefully framing a discreet photograph, it showed multiple full-length poses of a barely clad model. Instead of softly promoting the use of its product through a pamphlet or mail order follow-up, the full-page ad featured a coupon to order as many pink, white, or blue girdles as the customer wanted, "Individually packaged in SLIM silver tubes!"

ABRAM NATHANIEL SPANEL

The mastermind of the Playtex girdle's promotion, A. N. Spanel, was born in 1901 to a family of Jewish garment workers in Odessa. The young Spanel became a refugee at four, fleeing to France with his parents to escape anti-Semitic violence. In Paris, his father found work as a tailor while his mother washed clothes for a living. On the eve of World War I, the family received permission to immigrate to Rochester, New York.

In the United States, Spanel quickly tired of the classroom. In 1922, he borrowed money to invest in his first major invention, the "Vacuumizer" clothes storage system. With electric vacuum cleaners a new appearance in middle-class homes, Spanel conceived and patented a "Vacuumizer" clothes storage bag that relied on the new appliance. Instead of marketing directly to consumers, the canny inventor promoted his patented system to vacuum-cleaner manufacturers themselves, who would then include a clothes storage bag as a bonus in the house-to-house sales of "integrated" cleaning systems. Spanel never finished college, but was a millionaire by 1927. While the rest of the country entered the Great Depression, Spanel used his Vacuumizer proceeds to research his next project. As the country emerged from the Depression, the garment industry was one of those least affected by the downturn, and it was into this market that Spanel again sought entry, exploiting new rubber technologies researched during the "Vacuumizer" venture.

A BRIEF HISTORY OF RUBBER

Rubber, or latex,[4] was by the late nineteenth century an established commodity but was harvested exclusively from wild *Hevea brasiliensis* trees in the Brazilian Amazon until the 1880s. The Englishman Henry Wickham successfully escaped from Brazil with a store of the plant's seed in 1876 (leading to Brazilian litigation and British knighthood) and the plant's cultivation followed in Singapore, Ceylon and, most successfully, Malaysia, where the local climate allowed year-round harvesting and sustained few natural parasites for the tree.

Despite Wickham's theft, Brazilian harvesting provided most of world production into the 1900s. Then with the invention of the pneumatic tire in 1880 and, more importantly, the consumer automobile tire in the 1890s, demand for rubber rose precipitously, driving prices to more than $100,000 per ton (in 2008 U.S. dollars) on the London market by 1911. British industry invested massively in Malaysian production, and when the introduction of radial and retreading tires reduced demand, prices dropped to half a percent of their highest value by 1932—one of the first commodity bubbles of the new century.[5]

One effect of bursting rubber prices was to encourage entrepreneurs to find new uses for the now-inexpensive resource. Changes in its production further aided material innovation; while initially exported in hard bricks, latex was kept liquid with ammonia from the 1920s onward. When Spanel started the Vacuumizer Corporation in 1923, one of the main expenses in production was "rubberizing" the fabric of the bag with a thin film of the expensive substance. By 1932, the material was almost cheaper than the fabric it coated. In response, Spanel's new company marketed latex textiles. Initially, he turned the waterproof, elastic nature of the fabric into an asset by marketing children's wear—not only the "Play-tex panties" (which promised to "keep your child 'socially acceptable'") but also bed sheets, bibs, and sunhats for infants. While his "Play-tex" brand name marketed to consumers, the name he chose for the company's official listing had grander ambitions—"The International Latex Corporation."

A broader effect of the cheap latex available on the world market throughout the 1920s and 1930s was to slow research into synthetic rubber. Natural latex—a defensive substance in the sap of latex-producing plants—contains miniature hydrocarbon strands, or isoprenes. The tangled embrace of these strands on contact with air produces the waterproof, flexible quality of finished rubber. It was not until 1930 that DuPont labs synthesized their first synthetic hydrocarbon chains that had similar, though more limited, properties. Marketed as "DuPrene" until 1937, the product finally found acceptance as the rebranded "neoprene." Since neoprene was three times the price of rubber, however, and still lacked equivalent flexibility, its use was reserved for limited applications where its flame and chemical resistance were seen as essential.

SPANEL, INC.

After initial success in children's wear, in 1937 Spanel marketed the first Playtex product for women, the "makeup apron" (a short, poncho-like covering to be worn while applying cosmetics). Anticipating expansion, he reorganized the firm's physical location as well. The headquarters shifted from Rochester to Manhattan, occupying a suite in the Empire State Building (derided at the time as the "Empty State" building for its available square footage).[6] The firm's factories were moved to union-free plants in Delaware and Pennsylvania; a Georgia plant would open after the war. By the late 1930s, Spanel won renown for innovations in advertising packaging—he would share a 1938 Irwin Wolf Trophy for marketing with Henry Dreyfuss—as well as the production of latex itself.

One of the reasons Spanel had initially focused on children's wear in his latex business was the material's proclivity to catastrophic failure. While latex was more durable than artificial substitutes, when a break occurred in the material it would spread rapidly along the grain of its manufacture. This was a potentially disastrous outcome where adult issues of modesty or comfort were involved. However, along with the new investments in production after 1937, Spanel developed and patented innovations in the shaping and curing of thin latex surfaces that allowed the one-piece molding of complex shapes—and confined punctures to small holes instead of disastrous splits. In 1940, the firm put its full resources—including the full-page advertisement in the front of *Life*—into the introduction of its first adult "foundation."

FASHION LINES

In 1938 *Fortune* magazine reported that the U.S. foundation garment industry was worth 65 million dollars. Though girdle makers had been "terribly frightened" by the seeming move away from corsets in the teens and twenties, the 1938 *Fortune* report explained that "the corset isn't dead and never had been." Other shapes and terms for the garments—including "girdle"—soon gained currency, and by the end of the 1930s fashions were moving toward a greater constriction of women's waists and hips. By the time of the 1939/40 winter collections, Mrs. Adam Gimbel, wife of the president of Saks Fifth Avenue, announced: "Small Waists and new corsets are coming back. … They are very beautiful but uncomfortable, but women will wear them. … We have had comfort for years, now we are going to be dignified."[7] (It was precisely these "new styles" that were shown causing women in the 1940 Spirella Corset advertisement to be "scared pink.")

It was to this fear of discomfort that A. N. Spanel's 1940 "Living Girdle" advertisement spoke. Within the context of a girdled silhouette (a "look so smart in today's clothes") Playtex emphasized a product that was not restrictive, but rather "as natural as your own lines, slimmed down." Apart from the traditional static pose shown by the model's only clothed appearance, the rest of her photographed exploits—sitting down, reclining, fastening a shoe, and even playing tennis—seem designed to deemphasize the girdle's appearance in favor of its seeming "invisibility" to the wearer.

The actual experience of wearing a solid rubber garment turned out to be far less forgiving. But, at least according to the advertisement, the "living girdle" is not worn; instead, it "lives and breathes with you," and even "becomes a part of you." More importantly, unlike its competition, the relatively inexpensive latex girdle need not be custom-fitted—the "all-way" latex stretch adapted, albeit claustrophobically, to any wearer's body.

Spanel's marketing strategy was ingenious. While not questioning the need for "foundations" for fashion, he emphasized that the new "patented" product was "not a corset at all, but a revolutionary, modern method of curve-control." Spanel promised women the stylish figure of a corset without its discomfort, and the strategy worked—151,262 girdles were sold from the *Life* advertisement's coupon alone. Playtex's factories rushed to complete orders, racing to produce not only enough girdles, but also the distinctive metal packaging that highlighted the product on department store floors.

BATTLE LINES

The December 7, 1941 Japanese attack on Pearl Harbor, Hawaii, would have a significant impact on International Latex's fortunes—but not so much as the Japanese invasion of British Malaysia on the following day. While the bombing of Pearl Harbor would mean that American industry would turn entirely to a war footing, the invasion of Malaysia, and Japanese takeover of rubber plantations there, would mean that the United States was cut off from the Allies' largest source of latex. Not only was ILC left without a market for its consumer products, it was left without the basic commodity on which it had based its business and innovation.

By all accounts, Abram Spanel—who felt a great debt to his adopted country—was far more stricken by his inability to help the United States in its time of need than he was by the potential fate of his own company. But with the cutoff in latex supply, his production lines ground quickly to a halt. His purchases of advertising, so central to the prior fortunes of the company, did not. As of January 29, 1942, ILC recommenced purchasing advertising space in major American newspapers, but in a manner quite different than other companies of the time. Most advertising during the war years consisted of companies attempting to keep their name in public consciousness through a period of extreme consumer propriety (toward the end of the war, indeed, many companies quite blatantly attempted to increase expectations and incentives for the return to consumer spending). By contrast, ILC devoted its advertising purchases to the reprint of topical editorials and news articles in national newspapers, and so provided political support to the Roosevelt administration throughout the conflict. (Spanel would continue to purchase newspaper advertising space for editorials until the 1970s.)

While Spanel's personal fortunes seem not to have been too negatively affected by the war— in 1942 he moved into Drumthwacket, the Princeton estate that he would later donate as the New Jersey Governor's mansion—ILC struggled. By 1944, contracts to produce life rafts and inflatable boats for the Navy had literally allowed the company to remain afloat, and the firm's wartime administration was moved to its Dover, Delaware plant (where production of military goods was centered). As the company returned to production in 1946, it was this experience that led Spanel to fund a continual government research branch based at the Dover plant.[8]

THE NEW LOOK ON ICE

As fashion curator Valerie Steele has noted, Christian Dior's "New Look" in postwar fashion was not a new look at all, but rather a reassuring return to fashion developments as they stood in the 1930s—as if "put on ice."[9] While pure consumer articles like cars, appliances, and even houses grew larger, more streamlined, and more individuated after the war, the lines of fashion, and women's bodies in particular, pinched inward. "We were emerging from a period of war, of uniforms, of women-soldiers built like boxers," Dior himself explained. "I drew women-flowers, soft shoulders, flowing busts, fine waists like liana and wide skirts like corolla." Against the gray of wartime, such a look was indeed revolutionary; against the strides of physical comfort and occupation made by women during the war, it was intensely nostalgic.

As the "New Look" continued into the postwar decade, Playtex's advertising increasingly emphasized the connection of its body-shaping products to contemporary fashion. "Paris Designers acclaim invisible Playtex girdle as perfect way to the 'figure of the 1950s,'" announced one advertisement.[10] Instead of the multiple static views of Playtex's prewar ads, the image of mobility and comfort was reinforced after the war by multiple-exposure images of women leaping, running, and spinning while wearing the girdle and an almost invisible flesh-toned bra. (In the mid-1950s, Playtex's own line of brassieres appeared.)

915 figure was really no figure at all. Straight up-and-down boned corset made women look bulgy. Clothes completed potato-sack effect.

1926 figure symbolized the "tubular twenties" with its straight, uncorseted figure. Boyish lines were unflattering to many women.

1931 saw a changing figure. Rigidly girdled, bias-skirted fashions were more feminine, but not exciting by today's standards.

1947 featured the padded-hip, full-skirted fashions and the famous "New Look," which is as dead today as last week's corsage.

PLAYTEX® PRESENTS THE "FIGURE OF THE 1950's"

A slim, supple, vital figure that only Playtex gives with such freedom

Radical changes in feminine fashions within the average American adult's memory have been changes in *foundations* even more than in fashions.

And the girdle that has helped bring about the most recent revolution in silhouette is the sensational PLAYTEX Girdle. Made of tree-grown latex, PLAYTEX combines amazing figure-slimming power with complete comfort and freedom of action.

Without a single seam, stitch or bone, PLAYTEX fits invisibly under the newest, narrowest fashions. Its all-way action-stretch smooths the line from waist to hips to thighs, as no other girdle ever has.

For your fashions of the 1950's—have the *figure* of the 1950's—a slim, young PLAYTEX figure.

JACQUES FATH,

world-renowned designer of fashions, expresses the "Fashion of the 1950's" in this dress designed exclusively for the American collection of Joseph Halpert. It is figure-fitting, willowy-slim with shorter skirts demanding trimmer hips — so easy to have with the Invisible PLAYTEX Girdle.

GIRDLE OF THE 1950's is PLAYTEX— at all department stores and specialty shops, coast to coast. In slim, silvery tube: Blossom Pink, Heavenly Blue, Gardenia White; extra small, small, medium, large.

PLAYTEX LIVING® PANTY GIRDLE	**$3.50**
PLAYTEX LIVING PANTY GIRDLE with garters	**$3.95**
PLAYTEX LIVING GARTER GIRDLE	**$3.95**
Extra large PLAYTEX LIVING GARTER GIRDLE	**$4.95**

HEARD ABOUT PINK-ICE?

It's the newest of the PLAYTEX Girdles— shimmering smooth, extra cool, light as a snowflake, fresh as a daisy, actually "breathes" with you... in SLIM, shimmering pink tubes... **$3.95 to $4.95.**

INTERNATIONAL LATEX CORPORATION
Playtex Park   C1950   Dover Del.

PLAYTEX GIVES YOU THE YOUNG LINES, THE SLIMNESS-WITH-FREEDOM, SO IMPORTANT TO YOUR 1950 FIGURE

9.2
Playtex advertisement, 1952.

Subsequent years of fashion involved not a softening of the new-look silhouette but, if anything, increasingly extreme versions of it. And the connection between the girdle and the delineations of fashion was not just implied but explicitly stated by designers such as Oleg Cassini, Robert Piguet, Pierre Balmain, and Marcel Rochas. In one advertisement, Piguet himself declared: "My designs *require* the figure of the 1950s. A figure you can have—with PLAYTEX!"

With the years, new "innovations"—from flocked lining to "pink ice" ventilation—were introduced into Playtex's line. By the mid-1950s, the company was marketing almost as many models as a car company, each contained in a different-color metal tube (including pink, silver, gold, and blue). The company's marketing strategies are credited as central to the larger shift in women's underwear from a custom-fitted specialty to its reinvention as a packaged, consumer item.[11] As well as producing the highly recognizable silver tube packaging, Playtex's metals plant in Delaware fabricated dedicated floor displays, complete with sizing charts and space for the full selection of packaged girdles, allowing the customer to purchase a girdle without a single fitting. The packaging and marketing of the girdle was at least as important to its success as the specific way it in turn packaged and shaped the customer's own body.

As the mediated postwar world extended into the moving images of television, so too did Playtex, sponsoring an afternoon fashion show—"Fashion Magic"—on CBS, and finding creative ways to depict its product on television. As he sought out innovation in advertising, Abram Spanel cultivated a friendship with Edward Bernays, the author of influential texts on public relations and advertising such as 1928's *Propaganda* and a famous 1947 article in the *Annals of the American Academy of Political and Social Science*, "The Engineering of Consent."

INTERNATIONAL PLAYTEX

A particular image used by Playtex during this era is worthy of note. While the decision was made in 1961 to send only U.S. military pilots into space, thus automatically ensuring male astronauts,[12] the female iconography of the era had its own airborne counterpart—the air hostess. Her paradoxical combination of professionalism, sexuality, and domesticity has itself been identified as exemplary of the era's conflicting desires—with a latex girdle as an essential foundation. According to flight attendant Mary Inman (who would later spearhead antidiscrimination suits on behalf of her colleagues), the use of restrictive garments by midcentury flight attendants was an unwritten requirement for employment.[13] Another airline employee of the era noted pinches on the rear—to check for girdles—as a regular part of daily "grooming checks." The resulting image of restraint crossed with international romance was in turn relied on by Playtex in their marketing. "She wore it in Paris, beneath the blazing sun," a 1957 Playtex advertisement begins (showing a uniformed attendant starting her day at an outdoor café). "She wore it in New York, when the day was done,"[14] claims a cut to a moonlit terrace (with our heroine in the arms of a handsome partner). Throughout the advertisement, the word "girdle" is never used, instead what was by then a synonym: "Playtex."

UNDER THE SURFACE

By the time Spanel sold a controlling interest in ILC to the Stanley Warner company in 1954 (a $15,000,000 transaction in which Spanel's post as chairman was retained) the Dover government research program had become self-supporting. The first government order filled in the Dover plant was for the steel and aluminum K-1 pressure helmet, used by the Navy and Air Force to deliver oxygen in conjunction with MC-1 partial pressure suits from 1952.

As recounted by ILC researcher Leonard ("Lenny") Sheperd, it was the awareness of developments in pressure suit technology that flowed from this contract that resulted in the Metals Division—renamed the "Specialty Products Division" in 1955—investigating pressure suit design as a whole from 1951.[15]

Yet Sheperd's own biography illustrates the incongruities still present in ILC's improvised government program. While he would later become Apollo Project Manager for ILC, Sheperd was recruited not from engineering school but from A. N. Spanel's Rochester living room, as a particularly gifted television repairman, in 1951.[16] Given his own experience, Spanel had little time for professional qualifications and was quick to admire an inventive facility that mirrored his own. And, according to Sheperd and others, it was the former television repairman himself who in 1951 first proposed a future role for the firm specifically focusing on necessary changes in pressure suit design to allow for work in space—as opposed to a backup system for high-altitude flying. As Sheperd told a NASA oral history interviewer in 1972: "Our objective was always to enable man to do useful work in outer space or in vacuum, while the objective of the pressure suit was to enable a man to survive, bring an aircraft down to a level at which the atmosphere would sustain life, and supply only the mobility necessary to operate the aircraft ... our objective was to enable man to perform tasks, work tasks, even walk around."[17]

While ILC did in fact attempt to procure full-pressure suit contracts from the Air Force and NASA, starting in 1954 with the Air Force's open call for proposals (see Layer 7), the firm focused on crafting greater mobility in pressurized garments than had been achieved by any suit producer—including even David Clark's invention of link-net to meet the practical requirements of mobility in the high-altitude X-15 suit in 1956.

CONVOLUTED ORIGINS

It was to this end that, starting in the early 1950s, Sheperd and his team in Dover developed a remarkable material innovation. Beginning with the tomato worm insight of the B. F. Goodrich X-H1 suit developed by Russell Colley (see Layer 7), ILC attempted to produce a material that would combine the bellows-like constant volume of the tomato worm with the restraint and bladder functions necessary to keep the suit inflated and mobile. This single-layer composite, which would become known as "convolute," molded a natural latex bladder together with embedded nylon restraints, and was further restricted by nylon webbing along its axis of motion. The flexible latex could bellow and compress to allow motion of the body, while the embedded nylon simultaneously ensured a minimal change in volume of the bladder as a whole.

9.3
The ILC competitor for
the X-15 suit.

Unlike the large-scale bellows of the B. F. Goodrich suit, the convolute could be manufactured as a mass-produced component, in a range of sizes and shapes, thus opening a whole gamut of new possibilities for the optimization and testing of pressure suit assemblies. And the expertise that made such a puncture-resistant, complex latex assembly possible was the same flexible know-how that produced the firm's best-selling girdle. While Sheperd and other engineers supervised the company's pressure suit development, skilled craftswomen were requisitioned from the firm's Pennsylvania garment plant to take part in the delicate process of assembling and manufacturing the experimental convolute.[18] Indeed, a crucial innovation turned out to be the composite of two materials used elsewhere in the firm—the latex of the girdle, to which neoprene was added, and the nylon tricot of the other lingerie products which the firm was now promoting. (The introduction of the "Living" Bra in 1954 was followed by the patented "Cross Your Heart" configuration in 1957.)

The thin nylon layer, around which the rubber convolute was molded, served to stem the flexibility of the rubber—so it would maintain a constant volume even as it moved—and hold thicker nylon elements in place during the molding process. The rings and webs further restricted the convolute's expansion. As the firm acquired experience in sewing together rubber assemblies and fabrics throughout its product line, a flow of seamstresses followed into the specialty division, assembling prototype suits from nylon, convolute, zippers and cable.

While the suits produced by ILC were clearly successful in their own right, it would be more than a decade before the research would lead to a production contract. A strong competitor to David Clark for the silvery X-15 suit in the mid-1950s, ILC lost the contract but gained a small stream of research dollars from the Air Force to keep their efforts going. When such funds ran thin, the company funded the effort internally, guided by Spanel's own patriotism and pride in potential service to government.

After unsuccessful attempts to secure contracts for the Mercury and Gemini missions, ILC was to have its first success in the suit contracting process in 1962, with the call for an Apollo suit that could not only support an astronaut inside his capsule, or even during a tethered spacewalk, but act as his sole means of life support while walking and doing work on the surface of the moon.

To produce the oxygen, telecommunications, and other support systems for the suit, ILC associated itself with a team of subcontractors, including Republic Aviation and Westinghouse Electric. The competitive evaluation process resulted in a phone call to Sheperd in his Dover, Delaware kitchen. In the gathering dusk, Richard S. Johnston, head of Crew Systems Division at NASA, reported: "Lenn, we like your proposal; we want you to come down and talk."

With great enthusiasm, the ILC team swept into Houston. Their feelings were dampened by the news that NASA wanted ILC to act only as a subcontractor to one of its would-be competitors for the suit contract, the aerospace conglomerate Hamilton Standard. While NASA had the greatest confidence in the physical performance of ILC's offering, they were unsure of the organizational abilities of what at the time was only a 50-person division in Dover. (By 1965, ILC would become the prime contractor for the Apollo suit, with Hamilton as a subcontractor, and the Specialty Products Division would grow to almost one thousand people.)

DIVISIONS AND CONCLUSIONS

In 1958, the Special Products Division had been renamed the "Industrial Products Division." With the arrival of the Apollo suit contract in 1962, it became the "Government and Industrial Products Division." Finally, in 1967, under the larger corporate umbrella of ILC's corporate parent, the Government and Industrial Division was split as a separate entity from the "Playtex" consumer brand—"ILC Industries of Dover, DE," versus the rechristened "International Playtex Corporation."

Beyond this eventual corporate distinction, how should we distinguish between Playtex and ILC, between girdle and space gear? Each piece of clothing—girdle and pressure garment—prepared its occupant for an extreme space, an extreme midcentury atmosphere. Each used the material pliability of rubber to both adapt to the moving complex reality of the body and allow the body to adapt to its environment. And in each case, the wearer of each garment battled against it; 1950s stewardesses flying in the tropics reported their shoes filling with sweat from the rubber around their hips, and lunar astronauts developed blackened fingernails from working against the inflated pressure of the suit. Each, in its own way, acted as the literal foundation to a visual icon—whether the jet-age, New Look female of the 1950s or the space-age astronaut male of the 1960s. And, to each icon, not wearing the garment would have appeared impossible.

LATEX DEPARTURES

"Men were not meant to design for women. Men make clothes in which one can't move." With such fighting words, in 1954, the 71-year-old Gabrielle Bonheur "Coco" Chanel relaunched her fashion house, dormant since 1939. Chanel's success in the late 1950s revived the fitted but flexible silhouette she had first proposed with suit-and-skirt ensembles in the 1920s and 1930s, and by the 1960s it was this look, not the restrictive wasp waist and girdle, that dressed first ladies and fashionistas alike. The rest of the space-age decade would see new, flexible uniforms for flight attendants, and a self-conscious move on the part of youth culture away from the fashion system altogether.

But latex retained a starring role in one drama. In photographs of the lunar surface, ILC's latex is visible on only one part of the final Apollo suit—the fingertips. There, the need for tactile feedback, agility, and precision meant that, instead of the more protective cloth that covered the rest of the suit, the gloves appeared to show the spaceman's innermost pressure bladder extending out so he might properly touch the world. (In fact, insulation demanded two layers of rubber at the glove's extremities.)

It is because of such simultaneous flexibility and precision that the latex spacesuit has stayed with us. While the composites used have moved away from natural rubber toward laboratory-made rubber, EVA pressure suits remain flexible, inflated bladders—despite all proposals to the contrary. With a single exception (see Layer 17), even alternative "hard" pressure suits developed throughout the remainder of the century relied on Apollo's latex glove design as the most flexible and responsive to the body, rubber fingertips and all.

Notes

1. Christian Dior, quoted in Richard Martin and Harold Koda, *Infra-Apparel* (New York: Metropolitan Museum of Art, 1993), 21; from Valerie Steele, *The Corset: A Cultural History* (New Haven: Yale University Press, 2001).

2. Robert Latou Dickinson, "The Corset: Questions of Pressure and Displacement," *New York Medical Journal* (1887): 507–516; referenced in Jane Farrell-Beck and Colleen Gau, *Uplift: The Bra in America* (Philadelphia: University of Pennsylvania Press, 2002).

3. Dated by the *Oxford English Dictionary* to 1927.

4. While Playtex advertising would sometimes seek to distinguish between "natural" latex and rubber, the two are synonymous. One of the company's early ads had made the most extremeof such claims, saying products were "made of purest latex (not rubber)." The artificial distinction in this case brought a complaint from the Federal Trade Commission, which pointed out that "in truth and fact the products are composed of material consisting essentially of rubber hydrocarbon." Later advertisements were more subtle in their claims, and the first experiments in laboratory-made rubber appeared only after World War II.

5. Zephyr Frank and Aldo Musacchio, "The International Natural Rubber Market, 1870–1930," EH.Net Encyclopedia, Economic History Association, ed. Robert Whaples, December 19, 2002, <http://eh.net/encyclopedia/article/frank.international.rubber.market> (accessed October 25, 2008).

6. "Which is greater, the Empire State Building, sometimes called the Empty State Building, or the mind of the architect who conceived it?," *New York Times*, May 18, 1936, 13.

7. Steele, *The Corset: A Cultural History*, 156.

8. Interview by the author with Homer Rheim, ILC Apollo Program manager and past president, ILC Dover, October 23, 2003.

9. Valerie Steele, *Fifty Years of Fashion: New Look to Now* (New Haven: Yale University Press, 1997), 11–13.

10. Playtex "Living Girdle" advertisement, *New York Times*, February 12, 1950, 149.

11. "How it All Began—Playtex," Sara Lee Intimate Apparel Website, <www.balinet.com/history_ playtex.html> (accessed January 26, 2006).

12. See Margaret A. Weitekamp, *Right Stuff, Wrong Sex: America's First Women in Space Program: Gender Relations in the American Experience* (Baltimore: Johns Hopkins University Press, 2004).

13. Mary Patricia Laffey Inman, "Title VII and Equal Pay in the Flight Attendant Profession," Flight Attendant Symposium, San Francisco Airport Museums, November 30–December 2, 2004.

14. Playtex advertisement (videorecording), Ted Bates Historical Reel, Museum of Television and Radio, Paley Center for Media, New York.

15. National Aeronautics and Space Administration, "Interview with Mr. Mel Case, Senior Design Engineer, ILC Industries Inc., and Mr. Leonard Sheperd, Vice President of Engineering, ILC Industries," transcript, April 4, 1972, 20.

16. Born in Queens, Sheperd made an unlikely figure in Delaware. While fishermen at the nearby coast were initially skeptical of his approach to deep-sea casting (which involved attaching remote-sensing vacuum tubes to his line to record temperature and pressure),

they apparently in time came to seek Sheperd's counsel in their own efforts. Rheim interview with the author, October 23, 2003.

17. "Interview with Mr. Mel Case, Senior Design Engineer, ILC Industries Inc., and Mr. Leonard Sheperd, Vice President of Engineering, ILC Industries" (transcript, 24), NASA Oral History recorded April 4, 1972, courtesy of NASA History Office.

18. Roberta Pilkenton in discussion with the author, October 22, 2002, Dover, Delaware.

Layer **10** JFK

John F. Kennedy's face turns upward, collaged within the frame of a Mercury pressure helmet, the centerpiece of a 1962 painting by the British Pop artist Richard Hamilton. Various technological elements—conduits, knobs, and switches—float across the picture plane, divorced from their functional context. Kennedy stares outward, fixing his gaze toward an uncertain void that visibly weighs upon his shoulders. To his right, behind the helmet's outline, the cross section of a complex lens points away from his eyes. In view of the lens, at the very border of the frame, are printed the Cyrillic initials "CCCP."

Hamilton's painting is a palimpsest of the mythic and literal elements surrounding Kennedy's brief presidency. With its newspaper president in a hero's helmet, the painting rejoices in its seeming contradictions. The work's own title quixotically proclaims: *Towards a Definitive Statement on the Coming Trends in Men's Wear and Accessories (a) Together Let Us Explore the Stars.*

Along with that of his (mostly American) colleagues in the pop art movement, Hamilton's painting responded to the seismic effects of television and print media on global culture. Beneath its ironic surface, *Towards a Definitive Statement ...* contains essential truths about Kennedy's unique physical, cultural, and political situation. And with its complexity, it clarifies the extent to which these ostensibly separate elements blurred together to grant Kennedy a unique understanding of the impact of Yuri Gagarin's 1961 orbit. In the introductory essay to a catalog containing the portrait, the painter explains his effort to "fabricate a new image of art ... [for] man's changing state and the ... modifying channels through which his perception of the world is attained."[1]

SIGNALS AND SYMPTOMS

John F. Kennedy's television mythology traditionally begins with the 1960 presidential campaign and his masterful defeat of Richard Nixon in televised debate. Kennedy's consciousness of his own image and his unyielding control over it, however, had much deeper roots. Grounded in the fragile tissue of Kennedy's sick body, and extending to a consummate understanding of his own surface fashioning and presentation, the mastery of public media that Kennedy brought to American politics had a very private origin.

"[He] hasn't got a year to live," Sir Daniel Davis, a prominent British physician, told Pamela Churchill (Winston's daughter-in-law) after she brought her friend John F. Kennedy to visit him on September 21, 1947.[2] Alarmed by deficiencies in Kennedy's adrenal system (which allows the body to respond to trauma, illness, and emergency), Davis diagnosed Addison's disease—for which he knew no cure. The physician was at the time unaware of a treatment first developed in 1939, and perfected after 1949, which included pills and daily injections. Most radically, pellets of desoxycorticosterone acetate were implanted in a patient's thighs or back, releasing chemicals that could sustain life for up to three months. Kennedy first received this treatment during a stay at the Mayo clinic in 1948, and he would undergo the same surgery every 90 days for the rest of his life.

The exact details of Kennedy's illness and treatment are relevant to this discussion for several reasons. John F. Kennedy committed the United States to the enormous investment of the space program for reasons of national and global image. His own consciousness of image, however, began not with the advent of television news in the 1950s, but with the constant and expanding gulf that existed between the physical pain and illness of his private life and his public role as a socialite, soldier, congressman, and senator. Of further relevance is the fact that Kennedy's medical treatment—and the debates surrounding it in the White House—would themselves mirror the medical progression of the space race, as in the movement from invasive, chemical solutions to Kennedy's illnesses at the start of his presidential term to the focus on physical training and stamina that would hold sway by the time of his 1963 assassination. Of final significance is the influential condition of Kennedy's own body, from the disconcerting effects of his illness and treatment on his physical and psychological persona to the constantly fashioned and adjusted facade of clothes, hairstyles, and accessories that represent the iconic JFK.

MEASURING ME FOR A COFFIN

From the time he was an infant, Kennedy's family did not publicly acknowledge his illnesses, chiefly out of fear that public knowledge of his frailty would interfere with his—and their—ambition. His father, Joseph P. Kennedy—a liquor magnate and millionaire who by 1939 was U.S. ambassador to Great Britain—had an ambition for his male children that bordered on obsession. When, as a young child, JFK expressed interest in becoming a priest, his father dismissed any notion of a reclusive existence, remarking that he looked forward to having the Pope in the family.[3]

As a student, Kennedy was spirited away to private hospitals for extended stays. As a young man, he was able to enter military service (deemed a career necessity) only after his father intervened to prevent him from receiving a medical exam.[4] Throughout his life, his public persona existed as a deliberately constructed facade concealing a fundamentally damaged physique. "Took a peak [sic] at my charts again yesterday and could see they were mentally measuring me for a coffin!" Kennedy wrote from the hospital in January 1936 to one of his few confidants.[5] Considering flying in bad weather in 1948, he quipped: "It's OK for someone with my life expectancy," but suggested his friends take the train.

Beyond Addison's disease, a survey of Kennedy's fragile body would include an inflamed spinal cord, frequent and intense fevers, venereal disease, an intolerant stomach, deafness in his right ear, and a wide range of allergies that sometimes produced incapacitation.[6]

Reflecting the spirit of the times, Kennedy's treatment during his early career focused on a profusion of chemical solutions. Beyond the continuous dosage of medication for his Addison's disease, he received a range of daily injections for pain and exhaustion. From the time of his first congressional campaign, Kennedy had been treated regularly by Dr. Janet Travel, a socialite anesthesiologist who injected Novocaine deep into his back, sometimes several times a day.

By the time of his presidential campaign,[7] Kennedy was receiving additional injections from a New York specialist, Dr. Max Jacobson. These included painkillers as well as amphetamines to boost the president's energy level. Jacobson's injections were considered experimental (he would lose his license in the 1970s for injecting non-FDA-approved substances), and Kennedy kept the treatments secret, even from his official doctors. When Robert Kennedy found out about Jacobson's injections, JFK dismissed his concerns quickly: "I don't care if it's Horse Piss. It works!"[8] Kennedy considered the treatments so valuable that he flew the doctor and his wife to Europe with him in 1961, as the only passengers on a secret flight accompanying Air Force One. By this time, in addition, Dr. Jacobson was supplying Jackie Kennedy with medication for depression, also at JFK's request.[9]

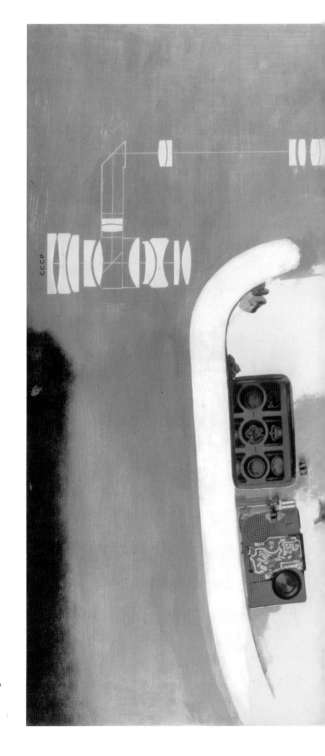

10.1
Richard Hamilton, *Towards a Definitive Statement
on the Coming Trends in Men's Wear and
Accessories (a) Together Let Us Explore the
Stars*, 1962.

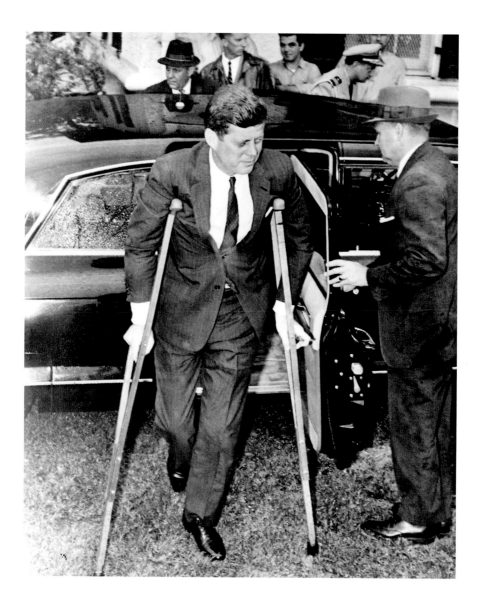

10.2

U.S. President John F. Kennedy walks on
crutches as he leaves his limousine to board the
presidential yacht *Honey Fitz* for a cruise
down the Potomac River with Japanese Prime
Minister Ikeda, June 21, 1961.

In his search for relief, Kennedy was also known to take drugs without a doctor or prescription involved. In May of 1961, Jackie Kennedy shared her concern with Dr. Jacobson over a vial of Demerol—an addictive and psychoactive painkiller—that she had found in her husband's bathroom. Her fears were confirmed—the drug could affect the president's thinking—and she arranged a transfer of the Secret Service agent who had been providing it.[10]

At this point, the president's official physician, Admiral George Burkey, took a radical step. On his own initiative, he contacted Dr. Hans Kraus, a renowned physical therapist, and called him to the White House. Dr. Kraus, known for training the Austrian Olympic Ski team, was sworn to secrecy and examined the president. The admiral's suspicions of Kennedy's overreliance on anesthetics and amphetamines were confirmed. The cocktail of medicines, along with a corset brace (that allowed Kennedy to stand without pain, but caused his muscles to atrophy), were taking a harsh toll on the president's body. "You will be crippled [by pain] if you do not exercise," Kraus told the president. Dr. Kraus was not the only one to perceive Kennedy's growing disability. By 1963, when out of the public eye, he was almost always supported by crutches and a cane, and his fate was the subject of a grim betting pool by the Secret Service, many of whom wagered he would be confined to a wheelchair during a second term.[11]

On Dr. Kraus's instructions, a special gymnasium was built adjacent to the White House swimming pool for the president to engage in a supervised series of exercises. Heating was installed to bring the pool's water to an inflammation-reducing 80 degrees. On hearing Kennedy's reluctance to have Kraus come and go from the White House to supervise the exercise (the president was afraid that the sighting of any doctor, and particularly the well-known Dr. Kraus, would allow the press to get wind of his physical problems), Dr. Kraus retorted: "It's your decision, but you will only get worse! What will they write about then?"[12]

While Kraus's exercises helped moderate Kennedy's growing disability, it is not clear whether any treatment could have prevented further substantial deterioration. By the time Kennedy was killed in 1963, he was spending more time than ever in the relatively pain-free environment of the heated pool; its temperature had crept up to 87 degrees.

During his presidential campaign and administration, Kennedy's efforts to obscure his sickness were direct. "Tell them I don't have Addison's disease," he instructed his press secretary, Pierre Salinger, when his then-opponent Lyndon Johnson was circulating rumors about his health. "Say I used to take cortisone but I don't take it any more." The secret cortisone treatments were, in fact, so necessary to sustain Kennedy's life that in 1947 his father traveled around the world secreting cortisone in safe deposit boxes, lest his son encounter a medical emergency while traveling.[13]

THAT'S NOT ME!

In retrospect, what is most unsettling about Kennedy's health problems is that his illnesses, and the drugs and devices used to treat them, had an undeniable effect on his physical self, as well as—it is hard not to believe—his execution of the presidency. The continuous pain he suffered forced him to limit his public appearances, and made him very conscious of the nature, duration, and effect of the physical appearances he did make. He would stay up after press conferences, watching the late-night replays and endlessly criticizing his own image. "Look at that camera angle!" he would say, "They're killing me!"[14]

Despite a superficial picture of health, with a seemingly permanent, glowing tan and resistance to illness, many of Kennedy's most distinctive physical attributes derived in part from side effects of his many treatments, and particularly his treatment for Addison's disease. Known side effects of the cortisone treatment included skin discoloration that resembled a "permanent tan," failure of the hair to turn gray, a swollen and sometimes puffy face (Kennedy had been heard to look in a mirror at his swollen face and say: "That's not me!").[15] In addition, an enhanced sense of confidence and power was frequently reported, as was an increased sexual desire,[16] and reduced resistance to infection.

While it is not clear whether Kennedy's illness contributed directly to his desire to construct and control every aspect of his public persona, it is clear that such a desire influenced much of his dealings with the press, and even his own inner circle. He would use his personal connection to reporters such as Hugh Sidey of *Time* to craft a comprehensive screen of personal details and information enhancing his image. "Let's make it 1200" words per minute, Kennedy told Sidey, regarding a piece he was going to print recounting the president's speed-reading ability.[17]

Kennedy was renowned, within the White House and beyond, for his immaculate personal appearance. This facade cracked only at the worst moments of crisis, whereupon it was a subject of major comment. The morning after the Bay of Pigs invasion became a fiasco, Kennedy shocked the West Wing by appearing with "disheveled" hair, and his tie "askew"—though it was also noted that he "sharpened up" by the time he reached the cabinet room.[18] Even during a crisis, John F. Kennedy changed his entire outfit up to four times a day, often going through six shirts alone. (The president owned scores of custom-made Italian and French suits, and favored a two-button cut that was then uncommon in the United States. His shirts were also custom made.) When his illnesses restricted his movement, Kennedy would dress and change clothes with the help of his valet.[19] Unique among modern presidents, Kennedy had a major influence on the style and dress of the country that surrounded him. A 1962 *Esquire* article on the new dress habits of the American male remarked: "The label on this man's shirt is 'JFK.'"[20]

Sometimes Kennedy's effect on male fashion was overstated. A Kennedy legend has the president single-handedly causing the decline of the American hat industry by failing to wear a hat to his 1961 inauguration. While this is inaccurate—Kennedy wore a top hat for all parts of the inauguration save his speech, restoring a formality to the occasion dropped by Eisenhower in 1953—Kennedy's subsequent bare-headedness, and the declining fortunes of American haberdashers, were enough to cause significant concern in the industry. A representative from the haberdashers' guild visited Kennedy in 1962 with a gift of several homburgs, as well as an entreaty to the president to use his style-making power to revive the hat-wearing habit. Kennedy's reply to the envoy is illuminating—he did not wear hats for any other reason except that they were extremely unflattering. To prove his point, he donned the proffered hats one by one. Even the hatmakers' representative, a witness recalled, admitted that they all looked "terrible."[21]

The inability of the president's head to elegantly support a hat was exploited by his public relations machine when, in the few days before the Cuban missile crisis became public, Kennedy

10.3

A mediagenic vignette: Kennedy leaving his Chicago hotel wearing a hat as part of the cover story surrounding his return to Washington during the then-secret Cuban missile crisis.

was rushed back to Washington to manage the crisis. The secretly sick president pretended to have an illness he did not have, and doffed one of the hats he usually did not wear to indicate his feigned ill health in the popular press.

WE'VE GOT TV

With his style and media expertise, Kennedy was able to control his own public image, and project his own image and ideals, further than any president before him. "President Kennedy and his advisers place boundless faith in his powers of persuasion on TV screens,"[22] *Time* remarked. The same article quoted an administration official: "We don't need the press anymore, we've got TV."[23] And yet, the same profusion of television media that made the control of Kennedy's image possible also created a media-driven world where a single event, if sufficiently captured and broadcast across the globe, could have complex consequences for the American presidency.

In the contemporary context of Vietnam, a single image such as the self-immolation of Buddhist monk Thich Quang Duc on June 12, 1962 (which highlighted South Vietnamese President Diem's repression of Buddhists) could radically shift American policy. On the very day he saw the image on America's front pages, Kennedy asked for the resignation of the ambassador to Vietnam. Several days later, he appointed political heavyweight Henry Cabot Lodge as a powerful replacement, a move seen by many historians as a turning point in the war's escalation.[24]

On May 3, 1963, Kennedy saw his administration's nonconfrontational attitude over civil rights called into question when an Associated Press photographer and a television camera caught sight of a 17-year-old African-American being attacked by Birmingham, Alabama police dogs. Accused of a civil rights policy that was more smoke than fire (watching the public spectacle of Glenn's launch on TV with the president a year earlier, Vice President Johnson had remarked: "If only he were a Negro!"), the president was moved to action when the images hit the airwaves and front pages, rushing to negotiate a truce in the Birmingham conflict.

At home and abroad, whether fighting the Cold War or managing domestic turmoil, Kennedy was waging a war not of tanks and territory but of instantly transmitted images and their powerful global consequences. Necessarily a creature of surface and image, he understood this transition and, when possible, embraced it in order to accomplish personal and national goals.

10.4

Thich Quang Duc, a Buddhist monk, burns himself to death on a Saigon street, June 11, 1963, to protest alleged persecution of Buddhists by the South Vietnamese government.

10.5

A 17-year-old African-American civil rights activist is attacked by police dogs during a demonstration in Birmingham, Alabama, May 3, 1963.

10.6
U.S. President John F. Kennedy, right, and Democratic
congressional leaders watch the launch of John Glenn
and the Friendship 7 Mercury capsule atop an Atlas
ICBM booster, February 20, 1962. From left: Rep. Hale
Boggs, House Speaker John McCormack (partially hidden),
Rep. Carl Albert, Sen. Hubert Humphrey, Vice President
Lyndon Johnson, and Kennedy.

POYEKHALI!

When it came to the Cold War, images were crucial—in both public and secret realms. While the fights over Cuba that produced conclusive evidence of Soviet missile installations were taken by the Eisenhower-era U2 spy plane,[25] the birth of satellite surveillance during the Kennedy administration was tremendously important in cutting through layers of uncertainty surrounding Cold War escalation. In August 1960, after twelve secret failures, the CIA managed to place in orbit its first spy satellite, the Corona.[26]

In January of that year, as Kennedy was inaugurated in Washington, the primitive satellite started ejecting cartridges of film taken over the Soviet Union into enormous nets trailed by specially equipped Air Force planes. The Corona satellite was designed to focus on reported missile installations—known because they had been the subject of boasts by Nikita Sergeyevich Khrushchev and other Soviet officials. The photographs dropped by the camera were initially suspected to be the result of a glitch—they showed pastures and wilderness—but they instead accurately confirmed that Khrushchev's talk of his own military strength, which had led candidate Kennedy to talk of a "missile gap," was just that—talk. Kennedy then understood that he had committed to one of the largest peacetime buildups of military strength (defense, space, and secret surveillance funding comprised more than 55 percent of his 1962 budget) against an enemy that he suddenly understood to have empty scabbards. To the chagrin of many inside the administration who saw America arming to meet an empty threat, the existence of the spy satellite was such a secret in itself that there was no suggestion that the newfound knowledge it brought should be shared with the public, or produce a shift in funding priorities.[27]

Thanks in no small part to the Corona satellite images, Kennedy knew that, militarily, the Soviet Union was a paper tiger (members of his administration had already begun talking of using an arms race to spend the Soviets into bankruptcy), yet it was clear that military posturing was one small battlefield in the larger propaganda landscape of the Cold War. Of special concern was the projection of success and efficiency to decolonizing and developing countries, seen as the most important fight of the Cold War. In this war, the image of Gagarin's voyage—down to his farewell to earth, "Poyekhali!," or "Off we go!"—was a dramatic success. As the invalid president contemplated the Hero of the Soviet Union, he saw an image that shifted the global balance.

Writing in his diary, British Prime Minister Harold Macmillan had characterized the difference between Kennedy's long-term strategic thinking ("not too strong") and his ability to act and think quickly about specifics ("remarkable"). Three weeks before Gagarin's launch, Kennedy told James Webb, the director of NASA, that he had decided against giving new funding to Project Apollo, resulting in an indefinite hold—Kennedy couldn't see the long-term use of the project.

Changing his mind in bed the morning of Gagarin's landing, Kennedy leapt into action, tripling the country's space budget within weeks, ordering schematic designs for every stage of the 1960s space program, and committing the country to landing a man on the moon by the end of the decade.

Later that same morning, pacing back and forth in the Oval Office (because of his back, the president sat down in his office only when protocol or the circumstances demanded he do so), Kennedy dictated a memo to Lyndon Johnson in his capacity as head of the National Space Council:

1. Do we have a chance of beating the Soviets by putting a laboratory in space, or by a trip around the moon, or by a rocket to land on the moon, or by a rocket to go to the moon and back with a man? Is there any other space program which promises dramatic results in which we could win?

2. How much additional would it cost?

3. Are we working twenty-four hours a day on existing programs? If not, why not? If not, will you make recommendations to me as to how work can be speeded up?

4. [A long question on booster technology.]

5. Are we making maximum effort? Are we achieving necessary results?[28]

Johnson spent the next week consulting with a collection of scientists and officials. In his answer to this last question, he wrote: "we are neither making maximum effort nor achieving maximum results necessary if this country is to reach a position of leadership."[29] In the rest of his response, Johnson recommended the increase of funds to allow 24-hour work on all space activities, and a further increase in the NASA and Department of Defense budgets to allow the achievement of "dramatic accomplishments." Chief among these accomplishments, Johnson recommended manned exploration of the moon as a goal with "great propaganda value."

"If we do not make the strong effort now," Johnson continued in his assessment, "the time will soon be reached when the margin of control over space and over men's minds through space accomplishments will have swung so far on the Russian side that we will not be able to catch up, let alone assume leadership."

As a result of this recommendation, the decision was taken by Kennedy to invest the country in a "Vastly accelerated space program, with the objective of landing a man on the moon."[30] While the decision was not announced publicly until a joint session of Congress on May 25, 1961, its content was made on April 28 and 29, seven days before the first astronaut would enter space.

As a result, any conception of the Mercury, Gemini, and Apollo programs as serial, or part of the progressive, steady expansion of activity by the American space program, is misconceived. The three programs were, at this point, entirely focused on one goal: landing a man on the moon before the Soviets and broadcasting the achievement. Only this spectacle would assure American supremacy—not so much in space, but in the minds and hearts of the impressionable world at home. The several advisors to Johnson credited in his memo include not only Wernher von Braun, the German rocket expert, but also Frank Stanton, president of CBS.

JFK's suspicion regarding the importance of a new kind of image, that of man in space, was confirmed a few days after Alan Shepard's suborbital flight on May 5, 1961. Habib Bourguiba, president of Tunisia, had arrived in Washington for an audience with the U.S. president. At a welcoming reception, Kennedy saw Bourguiba talking with Jerome Wiesner, the president's science advisor, and joined the conversation. "You know, we're having a terrible argument in the White House over whether we should put a man on the moon," Kennedy interjected. "Jerry here is against it. If I told you you'd get an extra billion dollars a year in foreign aid if I didn't do it, what would be your advice?"

"I wish I could tell you to put it in foreign aid," Bourguiba replied. "But I cannot."[31]

"Fill 'Er Up——I'm In A Race"

10.7

Herblock cartoon, May 1961. © The Herb Block
Foundation.

10.8
John Glenn shows President Kennedy a
pressure chamber and prototype Gemini space
glove, July 4, 1962.

Later that month, Kennedy began his second State of the Union Address of the year: "The Constitution imposes upon me the obligation to 'from time to time give Congress information of the State of the Union.' While this has traditionally been interpreted as an annual affair, this tradition has been broken in extraordinary times." The president paused. "These are extraordinary times."

Later in the speech, Kennedy committed the country to its most incredible Cold War adventure: "sending a man to the moon and bringing him back alive ... before the end of this decade."

From the perspective of Kennedy's knowledge of the media's power in the Cold War, the entire effort to go to the moon should be rightly understood as an elaborate apparatus for the production of a single television image. Kennedy approved plans to go to the moon because he—and perhaps peculiarly and particularly he—knew that the single image, however arduously achieved, could be magnified and extended globally, and, in an instant, change the world.

Notes

1.   One of only 20 paintings produced in the surrounding eight years by Hamilton, an attempt to "fabricate a new image of art to signify an understanding of man's changing state and the continually modifying channels through which his perception of the world is attained." From artist's introduction to the catalog for the exhibition "Richard Hamilton, Paintings etc. '56–64," October 20–November 20, 1964, Hanover Gallery, London.

2.   Richard Reeves, *President Kennedy: Profile of Power* (New York: Simon and Schuster, 1993), 43. Also Robert Dallek, *An Unfinished Life: John F. Kennedy, 1917–1963* (Boston: Little, Brown, 2003), 153.

3.   *The Kennedys*, produced by Elizabeth Deane, WGBH-TV, Boston, 1992 (PBS Video, Alexandria, VA).

4.   Reeves, *President Kennedy*, 44; Dallek, *An Unfinished Life*, 82.

5.   Dallek, *An Unfinished Life*, 78.

6.   Reeves, *President Kennedy*, 43; Dallek, *An Unfinished Life*, 398–399.

7.   Reeves, *President Kennedy*, 146; Dallek, *An Unfinished Life*, 398.

8.   Reeves, *President Kennedy*, 147; Dallek, *An Unfinished Life*, 399.

9.   Reeves, *President Kennedy*, 147; Dallek, *An Unfinished Life*, 399.

10.   Reeves, *President Kennedy*, 147.

11.   Reeves, *President Kennedy*, 242; Dallek, *An Unfinished Life*, 472.

12.   Reeves, *President Kennedy*, 243; Dallek, *An Unfinished Life*, 473.

13.   Reeves, *President Kennedy*, 42.

14.   Ibid., 326.

15.   Ibid., 243.

16.   Kennedy apparently regarded his sexual escapades as a form of physical relief. The British Prime Minister Harold Macmillan—happily married to his wife Dorothy for several decades at the time—recorded his discomfort when the topic of a summit meeting veered to his own sexual habits. "I wonder how it is for you, Harold?" Kennedy asked. "If I don't have a woman for three days I get terrible headaches." See Alistair Horne, *Harold Macmillan,* vol. 2, *1957–1986* (New York: Viking Press, 1989), 290.

17. Reeves, *President Kennedy*, 53.

18. Richard Reeves interview with Senator Albert Gore, in Reeves, *President Kennedy*, 95.

19. Reeves, *President Kennedy*, 314.

20. *Esquire* 57 (January 1962): 35–40.

21. Paul B. Fay, *The Pleasure of His Company* (New York: Harper and Row, 1966).

22. "President Kennedy and His Advisers," *Time*, June 22, 1962.

23. Ibid.

24. By October 31, 1962, Diem was dead in an American-sponsored coup d'état, organized in part by Lodge. See Stanley Karnow, *Vietnam: A History* (New York: Viking Press, 1983), 281–292.

25. The pilot of the Cuban U2 mission, Roger Chafee, would be killed as an astronaut in the Apollo 1 fire.

26. An official announcement boasted of a new "weather satellite," called Discovery 13.

27. Corona-Hilsman Oral History, John F. Kennedy Presidential Library; John Ranelagh, *Agency: The Rise and Decline of the CIA* (New York: Simon and Schuster, 1986), 324–326.

28. Memorandum from the Papers of President Kennedy, President's Office Files, Box 30, Special Correspondence, "Johnson, Lyndon B. 1/56-11/61," folder 62, John F. Kennedy Presidential Library.

29. Ibid., 64.

30. Ibid., 66.

31. Reeves, *President Kennedy*, 139.

Layer 11 CONTRACTUAL PHYSIOLOGY

11.1

The Multiple Axis Space Test Inertia Facility
(MASTIF), located in the Altitude Wind Tunnel at
NASA's Lewis Research Center (now the
Glenn Research Center).

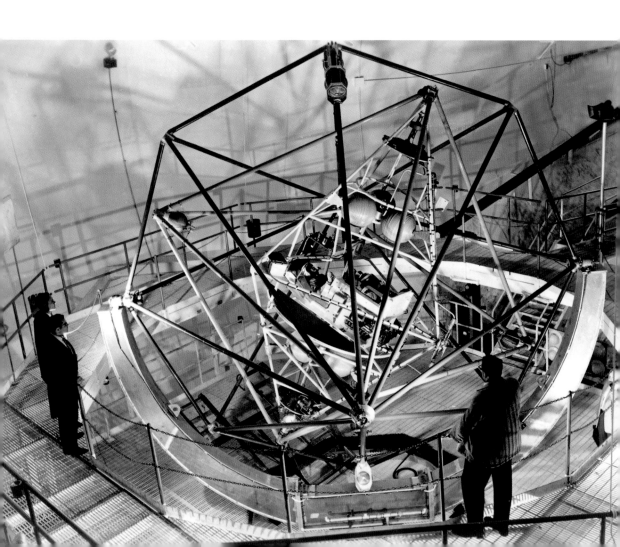

"Your body lurches against the tight harness that straps you to the contour couch. Then you rotate faster and faster. ... It is a wild, sickening sensation. Your vision blurs. The cold sweat erupts."[1] Writing in April 1960 in *Life* magazine,[2] astronaut Virgil "Gus" Grissom describes his firsthand experience in one of the "most impressive" devices used to train and test the Mercury astronauts: the MASTIF, or Multiple Axis Space Test Inertia Facility.

Grissom's words reinforce the popular idea that early astronauts were operating at or near the limits of human performance—both in their missions, and in the elaborate physical tests and simulations undergone in training. In ways that are both intriguing and essential to this narrative, however, this was never the case. Mercury missions did not approach or discover human limits. Rather, such biological limits had been researched and enumerated in obscure but astonishing tests throughout the previous decade. Once enumerated, these limits were literally inscribed into the network of contracts that constituted the military-industrial complex of the space race, in much the same manner as other technical limitations, for example gravitational, chemical, or aerodynamic, on the system as a whole.

As a result, in the course of an actual Mercury spaceflight—and so perforce in the "daunting" devices that sought to simulate the flight beforehand[3]—a Mercury astronaut's body was well within an envelope of performance that had been established years previously. So systematically were such measurements incorporated into rocket and capsule technology that the idea of an astronaut encountering circumstances on a mission beyond his body's own capacity was, at least in principle, impossible.

From the mid-1930s to the mid-1950s, a quantitative description of human frailty had emerged from research surrounding the physiology of flight. The research was conducted at the institutional margins of the Army and then the Air Force, more often than not within the stark horizon of the Southwest desert. During the most essential period of research following World War II, animals and then humans were subjected by the Air Force to ever-increasing tests of endurance. Within the desert's wide horizon, these isolated tests enumerated the operational limits of the human organism against speed, deceleration, depressurization, and altitude.

By the time of NASA's founding, these limits were available to engineers in government documents such as *Physical and Physiological Data for Bioastronautics*.[4] Thus, as NASA sought to quickly launch an American into space in order to meet the perceived Soviet threat, the problem was not, as the popular imagination had it, "machine-rating the men" (i.e., extending the abilities of men to let them ride rockets). Rather, it was "man-rating" the machines—adapting the technology of early ballistic missiles to operate within the boundaries of human performance. However, in the imagery of the time (and in the breathless space-age lingua franca of cyborg and superman), the immutable limits of the human organism were absent. Despite—or perhaps because of—the organic quality of the discipline, the essential history of aerospace medicine moved into the background of the space age.[5] Yet it is an essential story in that age's history.

SPACE IN ITS TOTAL FORM, 1

"[W]hat we call the upper atmosphere in the physical sense must be considered—in terms of biology—as space in its total form." So concluded the German Dr. Hubertus Strughold who, with his American counterpart Harry G. Armstrong, conducted foundational research into the physiology of flight starting in the 1930s.[6] One of Strughold's earliest experiments involved repeatedly anesthetizing his buttocks while at the controls of a biplane.[7]

A large part of the reason why research in flight physiology made its way so directly into the physiology of spaceflight relates to what Strughold was the first to articulate; for the human body, "space" as a limiting environment is encountered so quickly on ascent into the atmosphere that it can be reached in functional terms by what we now regard as primitive airplanes, let alone the missiles and rockets of the postwar era.

SPACE IN ITS TOTAL FORM, 2: THE DESERT

Frank Lloyd Wright described the Arizona desert as the place "where God is and man is not."[8] In the postwar era, the newly constituted Air Force chose the depopulated (and, after the Trinity atomic test, apocalyptic) landscape of the Southwest desert to test top-secret supersonic airplanes. With all its unearthly connotations, the desert would also be the place where the limits of man were established and enumerated.

Alongside pilots like Chuck Yeager and airplanes like the Supersonic Bell X-1, Air Force flight surgeons like John Paul Stapp Jr. migrated to the empty plains of Muroc (later Edwards) Air Force base to attend to the endless routines of flight tests. When, after successfully shepherding Yeager to exceed the sound barrier in 1947, Stapp was asked by his superiors to study the potential effects of even faster and higher flight on the body, he stayed at Muroc. (The desert, Stapp later admitted, shielded his experiments as much from the Air Force's financial supervision as it did from foreign espionage.)

From the detritus of the Muroc airfield, Stapp assembled a gunpowder-propelled sled, braked by a pool of water, in which to test the human response to acceleration and deceleration. As a test subject, he used himself, subjecting his own body to an increasingly violent set of maneuvers at many times the force of gravity. A continuous series of tests on his own endurance culminated in 1954, when, in a new facility in the gypsum sands of New Mexico, a rocket-powered sled accelerated Stapp to 663 miles an hour in five seconds, and brought him to a water-braked standstill one and a half seconds later.

During the test, Stapp's body was first pressed against his seat with 25 times the force of gravity, then thrust forward with almost twice as much force as he slowed. As the laconic Stapp reported, his vision became a "shimmering salmon … followed by a sensation in the eyes … somewhat like the extraction of a molar without an anesthetic."[9] In nine minutes, his vision started to return. The front surface of his body was badly bruised from pooling blood, and it was more than twenty-four hours before his nervous system displayed normal reflexes.

After this test, subsequent deceleration tests were performed on a shorter, air-powered track termed "Daisy." A fault in the "Daisy" mechanism meant that the officer taking over Stapp's role as test subject, Eli Beeding, was subjected to 83 G of deceleration, causing spinal bruising and temporary paralysis.

In still later work supervised by Stapp, "Project Manhigh," Colonel Joseph Kittinger flew the fragile monolith of a stratospheric weather balloon to 75,000 feet to test the limits of cold and

11.2

Major John P. Stapp on the Muroc deceleration sled, 1951.

11.3

A test on the Daisy sled in 1965.

altitude on the body. In his first jump, a sensor-equipped Colonel Kittinger was cast into a spin around the axis of his hips and, while falling 50,000 feet to the altitude at which his parachute opened, tumbled at up to 120 revolutions per minute. On his last jump, in August 1960, Kittinger experienced a free fall of more than 100,000 feet, bringing his partial-pressure-suit-clad body close to the speed of sound without (lasting) injury.

ARCHITECTURE OF THE DESERT

"In a landscape where nothing officially exists (otherwise it would not be 'desert'), absolutely anything becomes thinkable, and may consequently happen."[10] So wrote another architect, Reyner Banham, in his 1982 meditation on the American Southwest, *Scenes from America Deserta*. This is an analysis that holds true for the science-fiction tortures of medical testing in the desert, but (crucially) not for the physical bodies involved. While the laboratory architecture of jet sleds, deceleration troughs, and measurement prosthetics, produced by Stapp and his colleagues, was fantastic in scope, the fundamental architecture of human bodies at their core was emphatically everyday. Unlike the technological exploits of airplanes or spacecraft, whose dazzling innovations seemed as suited to the fantastic desert landscape as the streamlined trains of the prewar era, the biological bodies under the straps and trusses of the Air Force's medical testing were fragile, messy, and even fleeting.

Human subjects like Stapp and Beeding fared badly in the Air Force's tests, but animals fared worse; the 1952 death of a chimpanzee was "very messy."[11] Of the most disposable animals, like gerbils, it could sometimes only be determined that they "vanished into thin air."[12]

STRANGE TRAJECTORIES

The legacy of Stapp's work itself, together with the physiological component of space exploration, had a trajectory as limited as a rocket sled, first broadly covered by the media (Stapp made the cover of *Time* after his 25-G sled run), then referenced publicly (if disingenuously) in the recruitment and selection of male astronauts, then slowing before being braked assiduously by NASA officials concerned over calls to open astronaut selection to all those qualified physically for spaceflight. Most vexing to the NASA hierarchy were qualified women.

John Paul Stapp was the first to claim the broad relevance of his medical research in physical endurance. Contrasting his work on testing men with the elaborate mechanical testing that the rest of the desert base conducted, he noted: "There are only two models [male and female] of the human body currently available, with no immediate prospects of a new design; any findings in this research should provide permanent standards!"[13]

By 1959, however, NASA found itself (disingenuously) implying that the physical and psychological standards of spaceflight were met only by a select group of military men, and not by the population at large. As Stapp's own middle-aged frame makes clear in countless test photographs, even the first subject of the Air Force's test was more part of the mass of men than an elite of test pilots. And the first process proposed for astronaut selection seemed appropriately broad in its outlook. The draft of a 1958 advertisement for the Civil Service (i.e., nonmilitary) position "Research-Astronaut Candidate" sought only "(a) willingness to accept hazards … (b) capacity to tolerate rigorous and severe environmental conditions, and (c) ability to react adequately under conditions of stress." The advertisement added as a mere suggestion that "These three characteristics may have been demonstrated in connection with certain professional occupations such as test pilot, crew member of experimental submarine, or arctic or antarctic explorer. Or … parachute jumping or mountain climbing or deep sea diving … whether as occupation or sport."[14]

Contemplating the potential consequences of civilian astronauts, however, President Eisenhower ordered the advertisement scrapped. Preoccupied by the classified nature of the missile-related components of the Mercury vehicle (almost all of the rocket save the tiny capsule was a repurposed missile), the president ordered that only military personnel be considered. Publicly, however, the explanation for this choice stressed physical qualifications: "A qualified jet test pilot [a role only open to military fliers] appears best suited for the task."[15]

As a result, the popular image projected of the astronauts was specifically that they were qualified by special "physical and mental ability."[16] "Intensive physical and psychological tests,"[17] NASA administrator Keith Glennan announced in January of 1959, would be used to select final astronaut candidates from those who applied. Commenting after the selection of seven Mercury astronauts in April 1959, the *New York Times* explained that the final choice of America's first "Space Pilot" would by made by "surgeons," who would "choose the one of the seven judged to be at the peak of physical and psychological readiness."[18] This astronaut would be not only the better of his colleagues (it was implied), but the paragon of an entire nation.

In practice, however, the chief medical difficulty faced by NASA doctors was not in selecting between the various candidates interviewed, but rather in *distinguishing* between them. "The selection tests … were largely tests of tests," it was concluded, "conducted as much for their research value … as for determining any deficiencies of the group being examined." The selection of the Mercury Seven became instead a subjective judgment by committee. "We looked for real men," one member explained.[19]

LOVELACE WOMEN

The subcontractor to NASA for the Mercury selection tests was Dr. Randall Lovelace, a former flight surgeon who ran his own testing clinic in Albuquerque, New Mexico. Soon after finishing his tests on the Mercury men, Lovelace commenced a study of female responses to the same rigors. When he found that a group of female subjects could meet or exceed the same standards of stamina, balance, and endurance as the Mercury candidates, the self-described "First Lady Astronaut Trainees" gained a foothold in the popular imagination. By 1961, 13 women who had passed Lovelace's tests in Albuquerque gathered at the Naval Air Station in Pensacola, Florida for additional testing in jet aircraft. The tests were canceled, however, by executive order. Over a memo mentioning the women, Vice President Johnson had scrawled a vigorous response to NASA administration: "Let's stop this now!"[20] A congressional hearing followed in July of 1962, stage-managed by NASA to make quite a different point: rather than their physiques or innate capacities, it was the experience as military test pilots that best qualified the Mercury Seven as astronauts. "I am not 'anti' any particular group," Glenn explained in defending the male-only astronaut corps. "This is the cadre of people … who can best perform the functions of the program."[21]

11.4
One of Colonel Joseph Kittinger's record-breaking
jumps, recorded by an automatic camera on
his stratospheric balloon's gondola, December 1959.

11.5
Jerrie Cobb, FLAT, testing in the MASTIF, April 1960.

In a way that served both the prejudices and fears of NASA and the president, the vaunted physical qualifications of the Mercury astronauts were de-emphasized in favor of their "special" expertise.

LIMITS AND LAWS

At the same time as the physical qualification of astronauts was fading in the public eye, an understanding of fundamental physiological responses to the forces of spaceflight were permeating the interior technologies, procedures, and systems of nascent spacecraft. As technologies and systems designed for nuclear weapons transport were quickly adapted for human carriage, the biometric data codified by Stapp and others became a basic chapter of the NASA engineer's bible, both setting limits on some systems and forcing the invention of others.

In the end, biometric data were not used to extend human limits, or even to discover an exemplary astronaut specimen. Rather, such data were used conservatively—to ensure that both operational and training equipment for Mercury and later astronauts, MASTIF included, were built well within the physiological limits discovered in desert desolation. Compared to Joseph Kittinger's 120 rpm rotation, for instance, the MASTIF subjected astronauts to no more than 30 rpm in any direction.

In 1958, physiological testing by the Navy "cleared a roadblock" in the design of the Mercury capsule, when a prototype design for a contour couch, custom-molded to each astronaut's body, allowed Navy Lieutenant Carter C. Collins to endure a simulated G load of up to 20.7 G. The force corresponded to the load on a hypothetical Mercury capsule entering the atmosphere at an angle of 7.5 degrees. Engineers were "elated," as they projected the capsule experiencing less than half this force at its planned entry angle of 1.5 degrees, and so the contour couch became included in Mercury's design to ensure a "man-rating." The form-fitting contour couch provides the most literal example of the larger sense in which the technology of the early space age accommodated itself to its human subjects—and if not their literal bodies, then the enumerated limits of them.[22]

In 1951, Wernher von Braun declared: "The time has arrived for the medical investigation of the problems of manned rocket flight, for it will not be the engineering problems but rather the limits of the human frame that will make the final decision as to whether manned space flight will eventually become a reality."[23] Once such limits found their way into NASA's spacecraft design—as conservative margins of endurance, acceleration, spin, and temperature—they entered into not just the physical fabric of the spacecraft, but the organizational fabric as well. As part of a wider systems management approach of the sort pioneered by Schriever, Ramo, and Wooldridge for the original Atlas missile—the same rocket that would launch Glenn into orbit—the metrics of human endurance became imbedded in the contractual, cost-plus arrangements which governed the massive web of contractors, administrators, and manufacturers assembling NASA's launch systems across the entire United States. If only as a ghostlike outline of fragile limitations, the human was deeply imbedded in the organizational machine.

11.6
The MASTIF recorded in a time lapse photograph.

Notes

1. Virgil Grissom, "Wild 3-Way Tumbles in Mastif," *Life*, April 11, 1960, 58.

2. Loyd S. Swenson Jr., James M. Grimwood, and Charles C. Alexander, *This New Ocean: A History of Project Mercury* (Washington, D.C.: Scientific and Technical Information Division, Office of Technology Utilization, National Aeronautics and Space Administration, 1966), 244.

3. As described by Warren R. Young, *Life* science editor, "What It's Like to Fly into Space," *Life*, April 13, 1959, 133.

4. Captain Ellis R. Taylor, ed., *Physical and Physiological Data for Bioastronautics* (San Antonio: U.S. Air Force School of Aviation Medicine, 1958).

5. Maura Phillips Mackowski, *Testing the Limits: Aviation Medicine and the Origins of Manned Space Flight* (College Station: Texas A&M University Press, 2006).

6. Strughold's counterpart in American flight medicine was Dr. Harry G. Armstrong, who accorded his *métier* as cold comfort for complaints of airborne discomfort. In the late winter of 1933–1934, Armstrong flew as a medical observer from Detroit to Chicago in the passenger seat of an open-air biplane—a Berliner-Joyce P-16—that was, at the time, the military state of the art. Wearing the latest wool and leather flight clothing, the men were subjected to a steady −95° wind as the plane made headway through the thin high-altitude air. Armstrong's lips, face, fingers, and corneas all suffered frostbite. Untrained, he found himself flying the plane when the pilot's hands and arms lost the ability to move. "[S]omething has to be done about giving better protection to our airmen," Armstrong wrote his superiors on landing; "the unit responsible for development of that equipment … [should] be told to get busy and do something about this." Two weeks later, Armstrong received his orders to report to Wright Field and take charge of the work himself. See Mackowski, *Testing the Limits*, 15.

7. Historical Division, Office of Information Services, U.S. Air Force Missile Development Center Air Research and Development Command Holloman Air Force Base, New Mexico, *History of Research in Space Biology and Biodynamics at the Air Force Missile Development Center Holloman Air Force Base, New Mexico 1946–1958*, 33. After his death, Hubertus Strughold's reputation suffered greatly from revelations that linked him to medical experiments at Dachau during the Second World War.

8. Frank Lloyd Wright, *An Autobiography* (New York: Duell, Sloan and Pearce, 1943), 309. Wright claims the quote is from Victor Hugo, but he is presumably referring to the last line of Honoré de Balzac's story "A Passion in the Desert." "In the desert, you see, there is all—and yet nothing … God is there, and man is not." *La Comédie Humaine of Honoré de Balzac*, trans. Katharine Prescott Wormeley (Boston: Little, Brown, 1899), 406. The instability of the desert lodges even in the landscape of attribution.

9. John Paul Stapp, "Effects of Mechanical Force on Living Tissues I," *Journal of Aviation Medicine* 26 (August 1955): 286.

10. P. Reyner Banham, *Scenes in America Deserta* (Salt Lake City: Gibbs M. Smith, 1982), 44.

11. Mackowski, *Testing the Limits*, 149.

12. *History of Research in Space Biology and Biodynamics at the Air Force Missile Development Center Holloman Air Force Base*, 55.

13. John Paul Stapp, address delivered to the National Safety Forum, Fall 1955, quoted in *History of Research in Space Biology and Biodynamics at the Air Force Missile Development Center Holloman Air Force Base*, 46.

14. "Invitation to Apply for Position of Research Astronaut-Candidate," Announcement No. 1, NASA Project A, December 2, 1958.

15. Swenson, Grimwood, and Alexander, *This New Ocean*, 131.

16. "Seven Pilots Picked for Satellite Trips," *New York Times*, April 7, 1959, 1.

17. "110 Selected as Potential Pilots for Nation's First Space Flight," *New York Times*, January 28, 1959, 1.

18. *New York Times*, April 7, 1959, 19.

19. Swenson, Grimwood, and Alexander, *This New Ocean*, 163.

20. Stephanie Nolen, *Promised the Moon: The Untold Story of the First Women in the Space Race* (New York: Four Walls Eight Windows, 2002), 300.

21. Ibid., 247.

22. A final relevant human limitation was enumerated by the research of Joseph Stapp. One day in 1948, under exact circumstances that are unclear, a captain from Wright Field, Edward A. Murphy Jr., produced a set of deceleration gauges that he proposed be attached to the safety belts on Stapp's rocket sled to measure directly the force of the body. Each device could have been coupled in one of two ways; all were installed incorrectly and, after an exhausting sled run, provided no data. While it has been shown that the essential truth as well as versions of the phrase predate Stapp, by 1955 it had been popularly attributed to Stapp, and indelibly associated with the unfortunate Captain Murphy: "Anything that can go wrong, will go wrong." See Lloyd Mallan, *Men, Rockets and Space Rats* (New York: J. Messner, 1955), 101. For previous attributions, see Fred R. Shapiro, ed., *The Yale Book of Quotations* (New Haven: Yale University Press, 2006), 529. Shapiro quotes George Orwell in 1941: "If there is a wrong thing to do, it will be done, infallibly. One has come to believe in that as if it were a law of nature." George Orwell, *War-time Diary*, May 18, 1941.

23. Wernher von Braun, "Multi-Stage Rockets and Artificial Satellites," in John P. Marberger, ed., *Space Medicine: The Human Factor in Flights Beyond the Earth* (Champaign-Urbana: University of Illinois Press, 1951), 29.

Layer 12   SIMULATION

This book dwells on the intimate physical space surrounding the Apollo astronauts. If we expand our field of view only slightly, however, we encounter another novel and essential architectural context: digitally mediated environments of simulation and control. These are remarkable both for their contrast with the organic latex of the A7L suit, and for their equal relevance to contemporary architectural debates.

The ground-based simulators of Apollo, as well as the control environments that supervised and connected them, are in some ways the least public architectures of the Apollo era. In the current layer we will encounter spaces of simulation, including control rooms, which will receive further consideration in Layer 19.

A BRIEF PREHISTORY OF VIRTUAL SPACE

On February 9, 1934, President Franklin D. Roosevelt announced that he would be stripping the monopoly for airmail transportation from the gilded pockets of a few favored firms—the forebears of Northwest, United, and TWA—and transferring the duties to the U.S. Army Air Corps. Enduring thunderheads, whiteouts, and fog, the inexperienced Air Corps fell short, and literally fell out of the sky; in the few short months of the program, more than a dozen Army pilots crashed to their deaths.

To save resources, it was later revealed, the Air Forces had for some time prohibited practice flights outside sunny daytime hours, both to protect valuable airplanes and to save on fuel and maintenance. The limitations of the approach were drastically revealed by the airmail disaster; the Air Corps desperately needed a system to train young airmen for inclement conditions without putting their bodies, or expensive airplanes, at risk. Such a system was proposed in 1934 by the Link Company, originally founded as organ manufacturers in Binghamton, New York. (The mail routes, meanwhile, were quickly returned to the veteran pilots of reconstituted air carriers like United.)

PLANES AND PIPE ORGANS

The first patent given to the Link Piano and Organ Company in 1927 covered an improvement in the mechanisms of pianolas. The second patent, however, covered "an efficient aeronautical training aid—a novel, profitable amusement device," conceived by Edward Link, the firm's heir apparent and a frustrated pilot. It used the pneumatic suction and actuation long familiar to churchgoers in the form of organ notes to move, roll, and pivot a simulated aircraft fuselage.

By the early 1930s, the Link family had left the organ business, and opened a flight school using updated versions of its pneumatic training device. New versions incorporated instruments and the ability to black out the cockpit to simulate night flying. Against the background of the airmail scandal in 1934, Edwin Link, now an accomplished pilot trained on his own firm's machines, flew to Newark Field from Binghamton to meet with procurement officials for the Air Corps. A heavy fog blanketed the meadowlands, and the officers readied themselves to leave the airfield, assuming Link would be unable to land. When Link arrived regardless, his ensuing sales presentation was superfluous and his hosts from the Army immediately ordered six trainers.[1] By World War II, more than 10,000 of the ANT-18 basic instrument trainer, more commonly known as the "Link Trainer," were in operation, each constantly "tuned" to simulate a specific model of aircraft. Like the pianolas and organs in its own DNA, the Link Trainer operated through the analog exchange of air and electrical signals, the wavelengths and curves of the machine's adjustments and signals translating directly into motion of instrument, machine, and virtual pilot. The first set of adjustments to the 1929 trainer had taken Edwin Link six months; later machines still needed to be as regularly maintained as any other instrument in order to retain their crucial relationship to the characteristics of a specific aircraft.

12.1

The chassis of the Link Trainer shown here underwent subsequent modification by the National Advisory Committee for Aeronautics (NACA), precursor to NASA, at its Ames research center, in order to study analog electronic simulation of as-yet-unbuilt aircraft.

WHIRLWIND

In 1943, a Navy captain and engineer, Luis de Flores, put forward a different idea. Supervising the Navy's training facilities in Pensacola, Florida, de Flores proposed that instead of investing in more and more simulators to replicate different aircraft, the Navy should invest in a single machine that could be quickly set to simulate many aircraft—including, crucially, those that had yet to be designed. De Flores convinced his superiors, and the effort was housed (not coincidentally) within his alma mater, the Massachusetts Institute of Technology.

MIT's Project Whirlwind recruited a young engineer named Jay Forrester, and occupied its own secret and purpose-built facility on the Cambridge, Massachusetts campus from 1944 onward. While exchanging organ tubes for the sophisticated analog technology of new devices like the Harvard Mark 1 computer (whose 3,000 square feet of rotating drums were busy computing ballistic firing tables for the Army), the proposed device was superficially similar to existing trainers. When a pilot sat inside a blacked-out cockpit, he would press and turn airplane controls as if he were airborne. Most exciting to the designers of Whirlwind (and the Navy, which in 1944 invested $875,000 in a multiyear contract to build it) was the idea that, using the same system that would shift the device from one model of aircraft to another, an aircraft might actually be "simulated" that did not yet exist in physical reality.

The response times of programmable analog devices such as the Harvard Mark 1, however, were as yet inadequate for the dynamic simulation of real-life events, even as they remained extraordinary when compared to previous, human "computers." By the time a pilot had finished moving the simulator's joystick into a certain position, Whirlwind designers knew, the cabin and controls would already have to be tipping, moving, and twisting in response to his input. This could be accomplished by mechanical means for a single airplane—as in the design of gears and bellows for each model in existing simulators—but not in a way that could subsequently be altered for a new aircraft, and certainly not for a nonexistent one. Far more promising was the operation of the Army's newest computer, which, instead of processing data continuously using rotating drums and electrical contacts, used vacuum tubes to store and manipulate numbers as discontinuous states of "on" and "off." The Electronic Numerical Integrator and Computer, or ENIAC, had been designed at the University of Pennsylvania by Professors Presper Eckert and John Mauchly to replace the Harvard Mark 1 at the Army's Aberdeen Proving Grounds. It was the country's—and the world's—first digital computer. Unlike the simulator envisioned by Forrester, however, it could perform only different sorts of the same kinds of ballistic calculations, and lacked the flexibility to be fully reconfigured, as the "Whirlwind Computer" would have to be. Eckert and Mauchly's next computer, a "UNIVersal eniAC," would be designed precisely with such an ability in mind. For a brief time its acronym "UNIVAC" was synonymous with digital computing.

What followed transformed the Whirlwind Project, and laid the foundation for all digital computing that followed. The notion of a flight simulator was abandoned—in 1948 the airplane cockpit scheduled to be mated to a prototype Whirlwind Computer had been sent to a Cambridge scrap metal yard.[2] In its place, the ambitious Forrester focused his attention—and the government's money—on the creation of a multifunction, stored program computer. Like the new UNIVAC, the machine would be able to process numbers in different ways depending on what "software" was loaded into its dynamic memory (the so-called "von Neumann architecture" described by the Princeton mathematician in 1945).[3] Unlike UNIVAC, however, the Whirlwind was intended to process digital instructions in real time. While such a system had its origins in the need to process the real-time interaction of a simulator joystick and the aerodynamic and mechanical characteristics of a virtual aircraft, Forrester was soon proposing that the computer would be able to perform a whole range of battlefield tasks if funded adequately. Indeed, a 1947 presentation focused on the ability of a digital computer to perform the tasks of an antisubmarine "control room" of the sort established on destroyers and battleships during the Second World War.[4]

By the time Norbert Wiener published *Cybernetics* in 1948, Forrester's team was focusing on the technologies that would be reliable and instantaneous enough to make the Whirlwind into the device proposed by Wiener in his seminal text: "a mechanical brain."[5] The most novel of these technologies would form the basis for all subsequent computer memory.

Like its competitors until that time, UNIVAC used long toxic tubes of turbulent mercury to store information, their cycling waves a miniature surf of ones and zeroes, relayed electrically by sensitive microphones to the computer's processor. Apart from being messy and hot (the mercury had to be warmed to 200° C for the length of its wave cycle to match that of the computer's processor), the system could never approach the real-time response of which Forrester dreamed. To replace the mercury tubes, Forrester almost single-handedly invented a three-dimensional system of storage that would presage every subsequent piece of "random access memory" (RAM), and boosted the performance of Whirlwind past even the vacuum tube memory used by UNIVAC's other successors. Whereas other early computers marched through electronic data sequentially, as if they were still a stack of punched cards, Forrester's Whirlwind could respond interactively to inputs and outputs on the surface of a cathode ray tube.

By 1950 this work had completely replaced the Navy's original funding for the Whirlwind, and its new masters were those in the Air Force who sought to replace the country's aging radar defenses with a combination of humans and technology. The use of the word "system" to describe such an ensemble was unfamiliar enough to prompt the Air Force to provide a definition:

> The world itself is very general … [for instance] the "solar system" and the "nervous system," [etc.] … the Air defense system has points in common with many of these different kinds of systems. But it is also a member of a particular category of systems: the category of organisms … a structure composed of distinct parts so constituted that the functioning of the parts and their relations to one another is governed by their relation to the whole.[6]

From its role as a simulator, responsive and subservient to its human pilot, Whirlwind had transformed into an equal collaborator with its human operators in the context of the Air Defense system. This network would stretch in its final form—the Semi-Automatic Ground Environment, or SAGE—to offshore towers of radar sensors on the Atlantic, Pacific, and Gulf coasts, airfields and anti-aircraft missile installations across the country, and to displays, operations, and "control rooms" at 23 separate installations from Massachusetts to California.

And at the center of each installation was an updated version of the Whirlwind simulator, now known as the "Army/Navy Fixed Special Equipment," or ANFSQ7. The new "system" was seen as so symbiotic with its human operator that, as in an automobile, a cigarette lighter and ashtray were incorporated into its nuclear-missile control surface; Death, as it were, in two flavors.

From its origins as a simulator of the inhospitable, Whirlwind had also moved to become a handmaiden of the unimaginable. SAGE was never subject to a real Soviet bomber attack, but the network of control sensors and actuators (including human operators "in the loop") endured countless virtual ones. Prior to SAGE, war "games" had been costly affairs involving movements of scores of real troops and equipment to "simulate" ground and air combat. In its complex diminution of an oncoming bomber fleet to electronic signals, the Whirlwind was perforce a highly effective simulator of the type of attack it was designed for. As a result, its crews and facilities engaged in constant training, chasing virtual targets with light guns across a landscape no more real than the phosphors of their screens.

SPAM IN A CAN

As a newly formed NASA borrowed the technology of the nuclear arms race to fuel the new space race, one of the most important questions it faced was the role of human astronauts relative to the built-in control mechanisms of its repurposed nuclear missiles. In the first phases of the Mercury Program, the Atlas Rocket's origins as a self-guided ICBM, as well as concerns over human responses to weightless conditions, contributed to the so-called "pilot-astronaut" having little or no control over the trajectory of his craft. In his 1979 dramatization of the ensuing debate, *The Right Stuff*, Tom Wolfe has the heroic persona of Chuck Yeager, test pilot, deride the Mercury astronauts' subservience to control systems by dubbing them "Spam in a Can."[7]

In the subsequent Gemini and Apollo missions, a different philosophy of automation was developed. Starting with Gemini 3, NASA's orbital capsules were equipped with digital computers (Gemini's was made by IBM), designed to automate standard procedures, and, more importantly, provide astronauts with information from which to base rendezvous maneuvers and manually troubleshoot any departures from mission plans. (When piloting Gemini 13, Neil Armstrong and Dave Scott relied on information supplied by the onboard computer to resolve a near-fatal loss in capsule stability.)[8]

As described by Robert Chilton—an MIT systems engineer who would supervise control systems from Mercury to Apollo—the extensive and ultimately digital automation of the Gemini and Apollo spacecraft's normal operation was contrasted by the introduction of the astronaut "in[to] the loop" during any emergency, and especially during the most mission-critical procedures of orbital rendezvous and lunar landing.

12.2
SAGE equipment in Direction Center at McGuire
Air Force Base, New Jersey, Computer A, Magnetic
Core Memory Frame (magnetic core memory is in
the gridded structure of the two refrigerator-scaled
devices shown at center and right).

12.3

A 1963 US Air Force photo of a SAGE console in operation. From the original caption: "The heart of SAGE—electronic equipment coupled with man's ability to make decisions. Here, a USAF technician at a SAGE identification radar console selects tracks with a light gun for magnification and display on the Direction Center's summary board."

12.4

John Glenn in the Mercury simulator, Langley
Air Force Base, Norfolk, Virginia.

12.5

Gemini mission simulator, Manned Spacecraft Center,
Houston. Astronauts Ed White and James McDivitt
prepare to enter the simulator for extended mission
simulation.

This balance between automation and human problem resolution was decided for Apollo in its earliest stages, and clearly articulated at the summer 1960 conference hosted by NASA in order to familiarize potential contractors with the new program. As described by David "Davey" Hoag (another MIT alumnus who came from the Navy's Polaris missile program to Apollo), the strategy amounted to "Anything that the crew should do, you should have the crew do it, and not have [the] system automated"—especially in an emergency.[9]

As Chilton would reflect later, however, the decision "rose up and bit us," especially in its consequences for crew training, and the demands on mission simulation.[10] If astronauts, instead of mechanisms, were to be relied upon to respond to every conceivable crisis, then every such situation needed to be conceived and simulated in advance, well before the astronauts got off the ground. As described by Apollo 11 Command Module Pilot Michael Collins, electronic simulation became "the very heart and soul of the NASA system,"[11] and simulator availability came to control the entire schedule of 1960s astronaut training. And with the increase in reliance on the simulation devices came an increase in physical scale, and complexity to match.

DISPLAYING INFINITY

Project Mercury's only electronic simulator, one of three used by astronauts, relied on analog circuitry to simulate mission profiles and expose the astronaut to up to 276 anticipated "failures" in the capsules before he took flight. Like its 1950s flight trainer brethren, the electronic trainer was built by Link.

For Project Gemini, the number of mission simulators rose to four. A new trainer in Langley, Virginia (the home of astronaut training before Houston's Manned Spaceflight Center opened in 1964) used a pucklike craft atop a giant air table to allow astronauts to practice the spacecraft-docking skills Gemini was designed to test and develop. The "principal crew-training device"[12] for Gemini was a new electronic simulator built by Link. The two-man device used digital computers to simulate hundreds of additional potential failure modes, as well as mission profiles.

Mirroring the adjacent displays in the Houston Mission Control Center (to which it was, not incidentally, contiguous), the displays for the Gemini simulators are worthy of particular attention. Abandoning the simple closed-circuit television cameras of earlier devices (which projected dynamic images of models in response to simulated spacecraft motion), the Gemini simulators were the first to use what Link termed a "Virtual Image Display."[13] As with Mission Control's large television displays, the "virtual" display used a combination of transparent slides, combined with analog computer control, to generate images of target spacecraft for docking, stars, and the earth below. Together with a related "Infinity" visual system designed by Farrand (an optical specialist), the new system interleaved the projection of objects at different distances from the spacecraft using a system of mirrors and lenses. The result was the introduction of parallax, or the illusion of depth, into the spacecraft window display. Thanks to both the elaborate optics used, and the small angle of view out of the windows in spacecraft and simulator, the optical system meant that an astronaut moving his head inside the capsule would perceive a corresponding change in depth and angle outside the "virtual" window.

For the Apollo program (whose planning, we should recall, was parallel to and interwoven with Gemini), the number of types of simulators rose to eleven. Of these, the most important were the computerized simulators for the command module and lunar module. While Mercury and Gemini crews spent up to a third of their training in simulators, Apollo astronauts would spend more time confined behind "virtual" windows than they would behind real ones.

In Houston, two command module simulators and a single lunar mission simulator rose alongside Mission Control, their complex outlines shaped by the giant infinity displays, and rivers of wiring connecting them to mainframe computers, Mission Control displays, and cameras and models set up to replicate specific sites on the lunar surface.[14]

Further simulators focused on particular parts of the mission, like docking and landing. As with the Gemini simulator, their proximity to the electronic infrastructure of the Mission Control Center was deliberate. While early Gemini simulators had been built in Virginia, the cohabitation of capsule simulators and "real" control infrastructure was considered essential. Instead of being associated with the pilot training facilities of Langley, the new simulators were connected—spatially and electronically—to the infrastructure of Mission Control. While a Mercury astronaut might spend days in simulator training before a mission, Apollo crews would spend weeks and months inside the darkened, computerized capsules. But they were not isolated. In fact, they were connected to a collection of hardware and software that rivaled only the final launch infrastructure in its complexity.

Not only were the capsule simulators connected to the Mission Control Center, paralleling their "real" counterparts, they also connected to dozens of digital computers dedicated to simulating additional parts of the interconnected infrastructure of flight and launch. Modified IBM mainframes mirrored the performance of systems from the giant Saturn V to the tiny, state-of-the-art Apollo Guidance Computer—the first device ever to use (now-ubiquitous) multi-transistor microchips. This enormous virtual architecture was in many ways more important to the success of Apollo than the equally massive physical infrastructure of the event.

The architecture of simulation shaped Apollo's reality in new and unexpected ways. While the internal architecture of NASA's computers was transforming to continually process information, the astronauts were spending more of their time in data-driven landscapes. Even simulators that seemed "real," most remarkably the "flying bedstead," or Lunar Landing Research Vehicle (LLRV), designed to replicate the performance of the buglike lander under lunar gravity, contained electronic ghosts. The LLRV derived its responses to the astronaut's input not from physical connections, but from a computer (which counterweighted the pilot's body weight on the rear of the vehicle). The craft carried with it, digitally, an entire virtual moon.

12.6

The Gemini Visual Docking Simulator. Virtual display shows one of the unmanned Agena modules sent aloft with Gemini to allow astronauts to practice rendezvous and docking techniques prior to their use in Apollo.

12.7

A command module simulator (foreground) and Lunar Excursion Module simulator (background), Manned Spaceflight Center, Houston, 1967. A further command module simulator is in the far background.

LM MISSION SIMULATOR

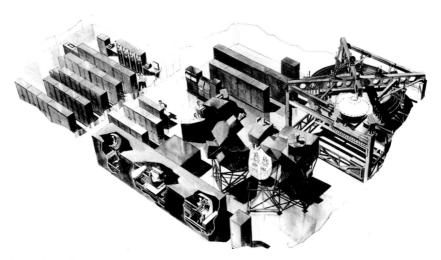

12.8

A rendering of the Lunar Mission Simulator, 1966.
At left are the computers driving the simulator,
and at far right and bottom are the collection of
models of the lunar surface and command module
used to simulate various portions of the mission.

12.9

A 1964 photograph of Lunar Landing Research Vehicle (LLRV) Number 1 in flight at Edwards Air Force Base. Note computer counterweight to pilot.

12.10

Digital displays from LEM Spaceflight Visual Simulator.

Like the multiscreen channels of Mission Control, the "Infinity" display systems of the major capsule simulators still used photographic transparencies and hand-painted models to create lunar landscapes, but the simulators' actual performance increasingly depended on the digital flow of information to simulate the unknown realities of lunar approach and landing. With the Apollo Program, Mission Control's IBM mainframes took over the task not just of simulating physical reality, but simulating digital reality as well. As a team from MIT and Raytheon raced to finish and test Apollo's silicon-based Guidance computer—the first of its kind—they wrote "emulation software" for the IBM 360 to replicate the performance of one computer with another, integrating their new device with mission simulations without tying up the valuable, experimental hardware itself.[15]

Finally, as Mission Control trainers sought better ways to simulate the unknown physics of a lunar landing, the final veil standing between astronaut and the digital fabric of simulation surrounding him was stripped away. Designed by General Electric (GE) with NASA, a 1964 system simulated lunar approach and landing without intervening transparencies and models, through the direct display of digital pixels. Limited initially to the depiction of a two-dimensional surface in perspective (though later extended to show relief and topography), the GE system was the first true, digital landscape of the space race, and, indeed, the first such digital landscape encountered by humankind. In its visual vocabulary of checks and crosses, the black-and-white cathode of the GE system is remarkable for its reduction of the lunar goal to its essence—a mastery not so much of earthly space or outer space, but of information space.

As we shall see below, this new, electric void would shape both the direct experience and the culture of the space age. And as we will observe in Layer 19, it would be only a short time before its rendered space would figure environments on earth: starting in 1966, the system was adapted for the simulation of an unbuilt, highway-based "cityscape" by Peter Kamnitzer and the Urban Laboratory Project of UCLA's School of Architecture and Planning, contributing to a new field of architectural representation. "We would like," remarked Kamnitzer in 1970, "to put the researcher, designer, decision-maker or the public at large in an environment where they could be exposed to a range of various futures. We will do this with computer simulation, which I believe will trigger the next creative leap in the human brain."[16]

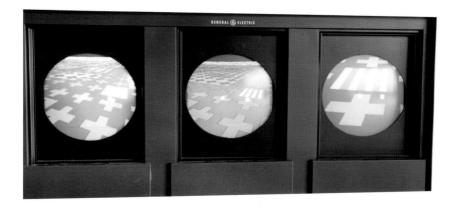

BETTER THAN THE SIMULATOR

"Throttle down. Better than the simulator," remarked Buzz Aldrin four days, six hours, 39 minutes and 37 seconds into the mission of Apollo 11, and just less than seven minutes before the *Eagle* touched down on the lunar surface.[17] The complex process of landing the Lunar Excursion Module, or LEM, on the moon's surface was the most uncertain part of the entire Apollo mission procedure. Correspondingly, it was the mission segment to which the most extensive simulations had been devoted. Out of 959 hours of training for Apollo 11, Neil Armstrong spent 285 hours in lunar landing simulators—not including time in the LLRV "flying bedstead." Aldrin spent more than a third of his own thousand hours of training simulating the landing, and each astronaut spent two weeks alone being briefed on these few, crucial minutes of the entire mission.[18]

The lunar module simulator in which Armstrong and Aldrin spent the bulk of their training time was in turn an object of enormous size; its glass optics outweighed the entire spacecraft whose interior it replicated. And yet the simulations it crafted were only as useful as the information that controlled them. While Buzz Aldrin commented on the accuracy of the landing simulations, errors in painting the canvas-and-plaster casts filmed by the LMS's closed-circuit cameras led a later astronaut, Apollo 15 commander Dave Scott, to become dangerously disoriented.[19] On landing on the moon, Scott reported, as if he had been there already, "a couple of shadowed craters, but not nearly as many as we were accustomed to."[20]

Despite Aldrin's nerves, Apollo 11 itself was not immune to failure. The accidental throwing of what had been an irrelevant dummy switch in the lunar module simulator led directly to a series of unexpected "program alarms" in the actual landing computer in the seconds before touchdown, risking the success of the entire Apollo enterprise. Yet the simulations were central to Apollo's legacy.

APOLLO, A PREHISTORY OF VIRTUAL SPACE

Viewed from the twenty-first century, the Apollo program can seem like a distant epic. Yet, recalling the manner in which epic tales have transformed the practice of storytelling itself over a much larger span, the media of Apollo, and in particular its simulations, have transformed the nature of contemporary reality as much as the fact of a lunar landing.

It is an accepted truth that the million or more semiconductors ordered for the computers of Apollo led directly to the birth of the modern computer era.[21] The ubiquitous simulations which those computers today produce present us with an irony. The very techniques designed so assiduously to protect and help transport the physical bodies of Apollo astronauts through the universe have directly contributed to the disembodiment of contemporary digital life.

CODA: FROM SPACE TO SPACEWAR!

When the programmers of the 1940s Whirlwind computer sought to demonstrate to visitors the machine's potential for real-time display and information processing, they did not turn to the elaborate test programs and information sets that would form the engineering basis for SAGE; rather, they used a deceptively simple demonstration program, Bouncing Ball. As visitors looked on, "A dot appeared at the top of the screen, fell to the bottom, and bounced, with a loud 'thok' from the console speaker. It bounced off the sides and the floor of the displayed box, gradually losing momentum until it hit the floor and quietly rolled off the screen through a hole in the bottom line." As simple as the program was, it would attract the attention of visitors for "hours."[22] It was the first digital simulation of a virtual environment. When the Whirlwind turned into Lincoln Labs' TX-0 (the direct antecedent of the SAGE computer), Bouncing Ball was replaced as a demonstrator by an interactive game, Mouse in the Maze. In the game, the same light pen used to track simulated bombers by the SAGE system was used to place cheese (or martinis) for consumption by a stylized (and, in the case of the martinis, steadily more erratic) virtual mouse as the rodent made its way through a maze also drawn by the "user"—now "player."

While Whirlwind and the TX-0 would transform in one incarnation into the centralized brains of SAGE and subsequent centralized simulators, another incarnation of Jay Forrester's monolithic Whirlwind would be the dispersed, decentralized and modifiable digital architecture of today. When IBM constructed its own mainframes, the (literally open) architecture of the first Whirlwind installation was covered up by the color-coded "facade" of commercial computation. These visual codes revealed purpose but disguised function and construction.[23] A clear contrast to IBM, in every sense, was provided by the MIT engineers who struck out from their Whirlwind experience to attempt a different approach to computing, founding the Digital Equipment Corporation (DEC) in an abandoned Maynard, Massachusetts textile mill in 1957. Whereas IBM only leased equipment to its customers and enforced a strict dress code for its employees, Digital was a quintessentially shirtsleeve, adaptive environment. Spreading out over the floor of the former Assabet mills, the company reasoned that since it lacked the sales force of an IBM, it would have to teach its customers how to modify and repair its computers, and gave away thousands of newsprint manuals and instructions for free. Whereas the internal "architecture" of the massive mainframes in Mission Control's Real Time Computing room still derived in part from vacuum tube technology, Digital's first product, the PDP-1, was based from scratch on the transistor, and introduced innovations that survive into personal computers of today. The machine spawned many elements of computer culture that are still recognizable. Digital's first customers included Bolt, Beranek and Newman (the Cambridge consulting firm that played a crucial role in the design of the Internet) as well as the computing lab of MIT.

Within MIT's already developed hacker culture (the first Whirlwind computer was by this point routing trains for MIT's Model Railroading Club), the Digital machine was far more attractive than staid mainframes—from which all modifications were barred. One of the first programs developed for the PDP-1 was a deliberate folly. Building on programs like Bouncing Ball and Mouse in the Maze, a team of students (many working under artificial intelligence pioneer Marvin Minsky) took it on themselves to provide a suitable demonstration program for the Digital machine. Raised on the pulp fiction of the postwar era (the team expressed a singular devotion to the work of pulp author E. E. Smith), the team decided quickly on the title for the program—Spacewar![24]

By 1961 hundreds of hours had been spent developing the game, which used control switches and improvised "joysticks" to move two spaceships into combat with each other around a gravitational "sun." The team encoded the entire night sky between 22.5°N and 22.5°S as a background to the conflict. And so in 1962, before the opening of Houston's Mission Control, an improbable competitor to the mantle of space simulation already existed—the first true videogame.

12.11
Nutting Associates, advertisement for Computer
Space, 1971. Note marketing by lingerie.

In 1965, when Digital shipped its latest replacement for the PDP-1, the PDP-8, Spacewar! was included in the machine's operating system as a demonstration. The PDP-8's low sales price (at $18,000, a fraction of the cost of a mainframe), its physical design (like a digital Pompidou Center, the PDP wore its user-modifiable circuits on its outside), and its size led the company to sell more than 50,000 machines.[25] A fashion analogy—the small, and even more popular miniskirt—led to the machine's nickname, the minicomputer.[26]

The influence of the PDP's built-in videogame on the virtual culture of the late twentieth century was enormous and direct. Engineering student Nolan Bushnell credited an encounter with the game on a Stanford University PDP-8 with his invention of the first freestanding arcade video-game—1971's Computer Space.[27] Like Spacewar!, Computer Space exploited the black depths of a phosphorous CRT to simulate the depths of outer space, and with it the exotic mechanics of zero-G maneuver. And yet while the subtleties of extraterrestrial dynamics seemed straight-forward to the programmers of Spacewar! and Computer Space, they were incomprehensible to a wider audience, and marketing strategies resorted to the use of diaphanously veiled models (wearing a more traditional Playtex garb than spacesuits). Bushnell decided he needed a game more comprehensible to a wider audience, and settled on a simulation of tennis. The small royal-ties from Computer Space would fund Bushnell's new company, Atari, whose tennis simulator Pong sold thousands of units as an arcade game. When in 1974 Atari was ported to a home version that could be attached to a television, sales were in the hundreds of thousands. It was only a short time before the background of outer space was again projected as what was now a videogame fantasy; Atari's Asteroids and Taito's Space Invaders held sway for the rest of the decade. As a result, by the time of 1982's Tron, virtual space had gone from a digital method for entering the unimaginable realities of interplanetary space to a playground in itself, in which the metaphor of outer space, so suited to the black voids and sparkling phosphors of cathode ray display, had itself become means of entry.

Notes

1. Link Corporate History, <www.link.com/history.html>, accessed July 4, 2007.

2. Kent C. Redmond and Thomas M. Smith, *Project Whirlwind: The History of a Pioneer Computer* (Bedford, MA: Digital Equipment Corporation, 1980), 60.

3. Paul Cerruzi, *The History of Modern Computing* (Cambridge, MA: MIT Press, 2003), 21.

4. Redmond and Smith, *Project Whirlwind*, 58.

5. Norbert Wiener, *Cybernetics: or, Control and Communication in the Animal and the Machine* (New York: Technology Press, John Wiley & Sons, 1948), 155.

6. This point is made in Thomas P. Hughes, *Rescuing Prometheus* (New York: Pantheon, 1998), 21, quoting Air Defense Systems Engineering Committee, "Air Defense System: ASDEC Final Report," October 24, 1950, MITRE Corporation Archives, Bedford, MA, 2–3.

7. Tom Wolfe, *The Right Stuff* (New York: Farrar, Straus and Giroux, 1979), 60.

8. David A. Mindell, *Digital Apollo: Human and Machine in Spaceflight* (Cambridge, MA: MIT Press, 2008), 87.

9. Ibid., 105.

10. Ibid., 92.

11. Michael Collins, *Carrying the Fire: An Astronaut's Journeys* (New York: Farrar, Straus and Giroux, 1974), 191.

12. C. H. Wooding et al., "Apollo Experience Report: Simulation of Manned Space Flight for Crew Training," NASA Technical Note TN D-7122 (March 1973), 48.

13. Myles and Ludwig Neuberger (Link Group), "Virtual Image Display for Space Flight Simulator," Technical Report of the Aerospace Medical Research Laboratories, Aerospace Medical Division, Air Force Systems Command, Wright-Patterson Air Force Base, Ohio, # AMRL-TR-66-58 (April 1966).

14. Mindell, *Digital Apollo*, 209.

15. As we will see, the use of the IBM 360 to simulate the functioning of the landing computer led indirectly to one of the major computing crises of Apollo 11's successful landing on the moon.

16. Gene Youngblood, *Expanded Cinema* (New York: Dutton, 1970), 250.

17. National Aeronautics and Space Administration, "Apollo 11 Technical Air-to-Ground Voice Transcription (Goss Net 1)" (Manned Spacecraft Center, Houston, July 1969), Tape 66/8 04:06:39:37, 313.

18. Robert Gilruth, "Flight Crew Training Summary" (memorandum), NASA Manned Spacecraft Center (July 15, 1969), Johnson Space Center Archives, University of Houston, Clear Lake, Box 071 Folder 52.

19. Mindell, *Digital Apollo*, 252.

20. David Scott, quoted in Training Office, Crew Training and Simulation Division, "Apollo 15 Technical Crew Debriefing," Report MSC-04561, Houston, TX, Manned Spacecraft Center (August 14, 1971, declassified February 10, 1978), 9–12; JSC Archive, University of Houston, Clear Lake, Cabinet Drawer 1.

21. Cerruzi, *The History of Modern Computing*, 188.

22. Van Burnham, ed., *Supercade: A Visual History of the Videogame Age 1971–1984* (Cambridge, MA: MIT Press, 2001), 44–45.

23. Reinhold Martin, *The Organizational Complex* (Cambridge, MA: MIT Press, 2005), 175–177.

24. J. M. Graetz, "The Origin of Spacewar!," in Van Burnham, 46.

25. Cerruzi, *The History of Modern Computing*, 128–129. Many thousand more PDP-8s would ship as single-chip emulators years later.

26. Cerruzi, *The History of Modern Computing*, 135.

27. Van Burnham, *Supercade*, 71.

Layer **13** THE MOON SUIT PLAYS
FOOTBALL

In the middle of 1965, during a six-week burst of creativity in which they developed the essential prototype of the Apollo A7L spacesuit that would go to the moon, a skeleton crew at the International Latex Corporation (ILC) found themselves breaking into their own labs and storerooms. Working 24-hour shifts, a dozen engineers and technicians climbed partitions and picked locks to obtain supplies and records while the rest of the company slept.[1] Some of the locked offices had been freshly vacated by visitors from the Hamilton Standard Corporation—previous supervisor and current competitor to ILC. It was a period in which ILC not only prepared the human body for an extreme environment, but also imbedded its own corporate functioning and identity in a new skin. This bureaucratic epidermis, as much as its layered spacesuit design, enabled the company to finally, and fully, incorporate itself into the Apollo program.

Several years after this flurry of activity, contemplating the subsequent decision to award the moon suit contract entirely to ILC/Playtex, an ILC engineer reflected on how much more defensible a choice of Hamilton Standard would have been to the many congressional committees supervising the space race: "How easy would it have been for [NASA crew systems director Richard Johnston] to stay with Hamilton Standard and say 'Gentlemen, we gave this contract to a Division of United Aircraft, one of the largest companies in the United States. What better judgment could I have shown?'"[2]

When NASA went looking for a contractor to make the Apollo lunar spacesuit, they sought not only the best design proposal, but an organization that would combine with and negotiate NASA's own complex bureaucracy. Although ILC's proposal had been the clear victor in NASA's evaluation of eight responses to an April 1962 Apollo "Extravehicular Mobility Unit Design and Performance Specification," NASA chose not to hire the firm directly, but rather as a subcontractor to Hamilton Standard. To Hamilton, the language of large-scale engineering contracts and relationships—reliability calculations, interface specifications, and above all the reams of paperwork deemed necessary to not only create change but communicate it across a vast institutional network—was a well-known lingua franca. To ILC, a company with some industrial experience but whose major income still came from latex underwear, it was a foreign tongue.

In asking Hamilton Standard to supervise suit production, NASA believed it was getting the best of both worlds. Hamilton would provide the suit's life support systems as well as the complex interface with NASA's organization. ILC, as subcontractor, would simply make the suit to NASA's specifications, as relayed to them through Hamilton Standard as the "prime" contractor. In other corners of the Apollo network, a structured relationship between two contractors had become a virtuous cycle of separation and solution. Even the mighty Saturn V itself was prepared in separate stages by separate companies (Boeing and Douglas), with points of contact—physically and organizationally—tightly controlled. The Lunar Rendezvous strategy similarly enabled the Apollo spacecraft to be separated into two distinct parts—the earth-adapted Command Module (CM) and the moon-adapted Lunar Excursion Module (LEM), whose problems and manufacture could in turn be separated into different constraints, committees, and corporations—North American Aviation and Grumman—in a setting of operational collaboration and friendly competition. But as applied to the Apollo spacesuit, such a strategy subjected the two organizations, ILC and Hamilton, to intolerable pressure. This pressure, combined with the complex demands placed on a single spacesuit prototype to serve as both intravehicular life support backup and protection for extravehicular excursions, produced not a virtuous cycle of innovation but a vicious one of failure and blame. It was a busy and bitter contest, and its final chapter was settled on the gridiron of a football field.

For all its difficulties, however, the "shotgun marriage"[3] with Hamilton Standard provided crucial training to ILC. During this period, ILC was able to not only master the procedures of large-scale engineering, deemed essential by NASA, but, perhaps just as importantly, also achieve crucial successes by abandoning procedure and precedent where necessary. The development of the Apollo suit from 1962 to 1967 is a story of choreographing softness and hardness, flexibility and implacability, permeability and resistance—material, cultural, and organizational. It is a story of hybridizing and buffering between materials and components of the suit itself on the one hand, and between the institutional and administrative nuances of its authoring organizations on the other. In this way, the many-layered suit becomes not only a map of the successive operational requirements to which competing versions were subjected in a relentless process of parameter setting and performance review, but also a literal embodiment of the corporate entities, private and public, that both established and were transformed by that process.

CLOTHING AND THE ORGANIZATIONAL BODY

In 1962, the first suits provided to NASA under the new Apollo suit contract were not collaborations, but rather additional versions of the garments ILC had successfully submitted for testing in the contract bid. Originating within ILC's small Special Products Division, these suits were designated model SPD-143.

Some twenty of the suits were produced and subsequently used by NASA, as well as other contractors like Republic Aviation, North American, and Grumman to start research on the suits' performance, and their interface with the wide range of Apollo's nascent subsystems. These interfaces ranged from the shape of acceleration couches in the Command Module to the size of the (crucial) exit door to the lunar lander.

Surviving versions of the SPD-143s show not only the Apollo suit's beginnings, but also the initial process of improving the design. The earliest unchanged models show ILC's original approach to the 1962 competition: a bladder constrained by alternate swatches of green ballistic nylon and molded latex convolute, arrayed around the body in relation to the jointed and fixed parts of the astronaut's anatomy. A further set of SPD-143 suits shows the garment itself becoming the test bed for analysis by both ILC and Hamilton Standard. Surviving examples show the suit becoming a home to all manner of experiments in cables, webbing, joints and restraints.

Suits surviving from the period immediately afterwards, from 1962 to1964, represent a collaborative effort by ILC and Hamilton to solve successive, and seemingly intractable, design problems. These suits were designated the A1H through A4H (A for Apollo, a number from 1 to 4 designating the suit generation, and H for Hamilton Standard, the official contractor). The increasingly bulky surface of this series of suits testifies to the growing complexities in design and organization faced by Hamilton and ILC, and to the challenging atmosphere their combined efforts faced.

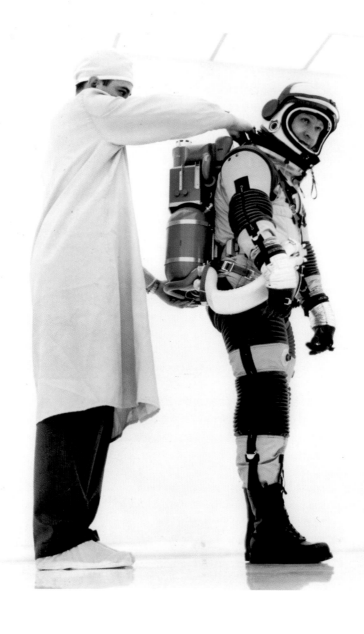

13.1

ILC Specialty Products Division (SPD)-143 under initial testing.

13.2

SPD used for experiments in thigh and hip restraints.

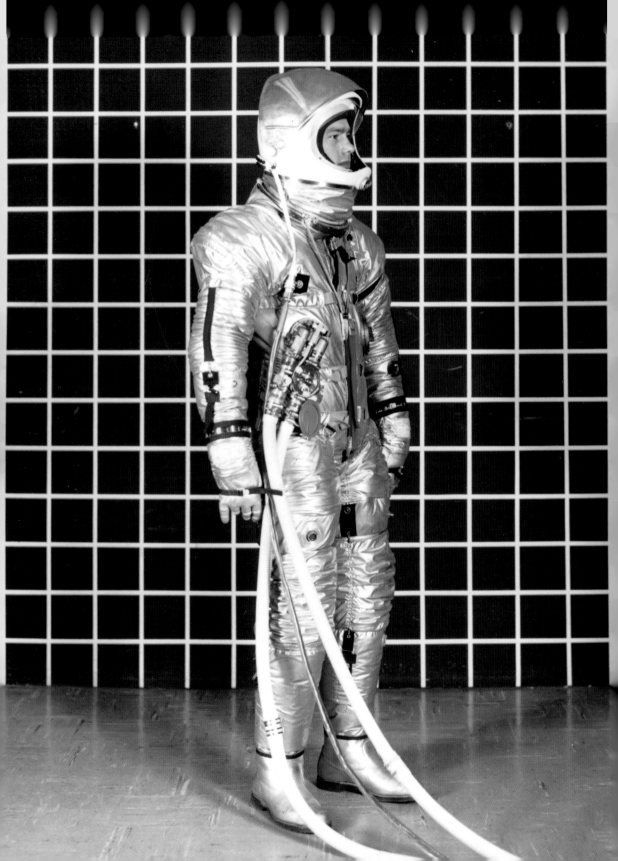

A significant number of the challenges offered by the lunar surface were not even addressed in this period. In the initial 1962 suit comparison, the so-called "Thermal Micrometeoroid Garment" (TMG) was provided by NASA, not by the suit contractors themselves. This garment, conceived as a separate outer layer over the pressure suit, was designed to protect Apollo astronauts from temperature extremes expected on the lunar surface, as well as a fine rain of microscopic meteoroids, which could otherwise puncture the suit, and whose precise quantity and velocity remained unknown.

What became increasingly clear during this period was that the functional requirements of the suit's operation outside the CM and LEM—its so-called "EVA," or extravehicular activity—directly undermined its functionality and performance for "IVA," or intravehicular, operations. The early decision to combine these two functions in a single suit had been a direct result of the stripped-down Lunar Rendezvous strategy deemed necessary to accomplish the landing before 1970. Nowhere was this compromise more clearly felt than at the interface of the lunar vehicle with the astronaut himself—the suit's surface.

To achieve necessary mobility in the suit's arms to allow for lunar sample collecting, for instance, the size and rigidity of the conical shoulder elements on successive AX-H suits increased. This, however, led to severe complications with the suit's fit inside the Command Module. An urgent telex from North American Aviation (contractor for the Command Module) outlined the problem in the dimensional language of engineering: "The minimum elbow breadth in the pressurized condition was 30.5 inches which is greater than the width of the couch—24.0 inches—and the requested minimum of 26.0 inches. The crew in inflated suits will not be able to get into the crew couches."[4]

13.3
AX2H suit from 1963.

13.4
One technician wearing an AX2H suit assists another into the Thermal Micrometeoroid Garment (TMG), mid-1963.

13.5
Simulated low-gravity testing on a mock
lunar landscape, Manned Spaceflight Center,
Houston.

When Command Module technicians attempted to inflate all three suits after test subjects had already secured themselves in the couches, the results were no better—the center astronaut broke free of his chair and "popped up like a piece of toast."[5]

Further problems arose. Under simulated testing in lunar gravity in November 1963, recumbent test subjects wearing AX2H suits were trapped on their back "like tortoises," leading NASA to issue a formal finding of contractual failure against Hamilton. In March of 1964, the next version of the collaborative suit, AX3H, was sent to Long Island for testing in Grumman's developing Lunar Module design. The results were so unsatisfactory that astronaut Gordon Cooper announced: "I would not go to the Moon in that suit"[6] (adding that he would choose the David Clark-designed Gemini suit instead, despite its many drawbacks in mobility and comfort). Indeed, the suit was sized so poorly to the Lunar Module that it was deemed "incompatible with the spacecraft."[7]

In April 1964, NASA announced that it was inviting David Clark to provide a revised Gemini suit for the first set of three Apollo flights (termed "Block 1"), in which equipment and rendezvous techniques would be tested without the need for a high-mobility EVA. This allowed more time for the Apollo suit design process, but also clearly revealed NASA's dissatisfaction with the state of those efforts.

In attempts to resolve an expanding set of failures, cultural differences between ILC and Hamilton became clear. To ILC engineers like Mel Case, Hamilton's requirements for suit development—such as the prohibition of screws below a certain diameter—seemed "not related to the space suit field, [but] the aircraft industry." As Case added, ILC "couldn't live with that."[8] And yet to Hamilton's highly trained engineers—students of the latest methods in testing, planning, and management—ILC's team practiced what they called "ragged art."[9]

In a set of continuous conflicts, ILC would produce a mechanism meeting the suit's functional requirements, only to be informed by Hamilton that a multiplication of each component's tolerance figures produced an unacceptable resulting risk. When informed of such a tolerance problem in the cable hardware restraining the suits, ILC precisely machined matched sets of components by hand, meeting the calculated requirements. But this only led to further howls of protest, as Hamilton engineers argued that the custom parts would make subsequent repairs and adjustments impossible. To the craftsmen of ILC, Hamilton engineers' approach to the human body in suit design closely resembled their portable life support system ("PLSS"), the backpack that a lunar astronaut would wear. The diagrams for this device reduced the human body to mathematical inputs and outputs on an engineering diagram, leaving him as just one "black box" in a systems-engineered cyborg.

ILC engineers disparaged in particular the Hamilton proposal for an additional backup bladder, wrapped around the primary latex pressure envelope. Such a proposal made mathematical sense (bringing a decreased risk of failure), but in tests thoroughly immobilized the astronaut, defeating the suit's original purpose.[10] (As recounted in Layer 8, such a double bladder would prove almost deadly for cosmonaut Alexei Leonov.) ILC's adaptive, experimental design process was in turn shocking to Hamilton. Measurable, verifiable, and repeatable procedures were the common speech of a military-industrial design culture, and they were a language in which Hamilton accurately judged ILC a novice. As a result of these clashes in culture, and the very real challenges the suit design process faced, the contract "progressed at high cost with very little accomplishment from 1962 to early 1965."[11]

13.6
The Hamilton Standard "tiger suit" in 1965 testing.

CHASING TIGERS

Under terms added to its contract in 1963, Hamilton was responsible not only for suit design and production, but also for producing an increasingly specific list of tests a successful suit would be able to pass. As design efforts ran further and further foul of these marks, Hamilton sought to eliminate what it saw as a substantial organizational variable—the presence of ILC on the suit contract. Hamilton had started internal helmet and joint designs in 1963, and in 1964 the organization hired Dr. Edwin Vail, a noted Air Force researcher, to head an internal "tiger team" of suit designers; the resulting "tiger suit" was completed in 1964. Competing against Hamilton's internal "tigers" for resources and attention only added to ILC's frustration and recalcitrance. The suit itself was a failure; a December 21, 1964 internal NASA memo concluded tersely: "Tiger suit a bust. Forget it."[12]

In February 1965, Hamilton Standard officials requested a meeting with Richard Johnston, head of Crew Systems for Apollo, to announce their decision to cancel ILC's subcontract for suit provision. They brought with them a "justification book" containing "copies of correspondence concerning overruns, delinquent hardware responses, delinquent qualification reports, delinquent reliability reports, delinquent development reports, examples of personality clashes, etc."[13] NASA officials agreed that it was Hamilton's prerogative to fire its own subcontractor,[14] and ILC received a telegram terminating its work on February 25, 1965.[15]

To fully understand the events that subsequently transpired—which would lead ILC from the wilderness into the heart of Apollo suit development—it is necessary to examine not only the landscape of procedure and performance surrounding the suit design process, but the social landscape surrounding it as well.

One of Hamilton's perceived criticisms of ILC's operations was that its engineers had graduated "only from the school of hard knocks." And, in a way, this was true. As previously noted, ILC's Apollo manager, "Lenny" Sheperd, was hired at ILC from a role repairing the televisions of its chief executive. Other top ILC executives, such as manager George Durney, were equally qualified—or unqualified— for high-technology production. Durney's mechanical skills—he was the son of a local auto dealer— saw him drafted into the World War II Army Air Forces as a mechanic; only a field promotion allowed him to finish the war as a pilot. Durney returned to his Delaware home for a job selling sewing machines at a local Sears & Roebuck. But his resulting mechanical ability with sewing machines, along with his qualifications as a pilot, brought him into the 12-person team that assembled ILC's first suit bid for the Air Force in 1955.[16] By 1962, Durney was Sheperd's lieutenant, managing day-to-day suit operations for ILC—including the deteriorating relationship with Hamilton.

ILC's Apollo efforts would always be managed by such "hard-knocks" engineers (indeed, "hard-knockser" became a title of respect inside the firm).[17] But the larger conflict between Hamilton and ILC accompanied an expansion of ILC's organization. From 1962 to 1965, the initial improvisational ILC team was augmented by a generation of young engineers like Mel Case, Al Gross, and Homer Rheim—each of whom would manage components of the final Apollo suit. Eventually becoming a self-described "buffer" between the "hard-knockers" and the larger Apollo effort, these young engineers were catholic in their appreciation both of ILC's free-form creativity and of the need for repeatable measurable results expressed by Hamilton and NASA. As ILC's special products division came to employ scores more engineers, the company introduced its own "precision management" techniques—forms, tests, and procedures—to track the expanding design process.

Further masked by the high-level conflict between ILC and Hamilton, and even by NASA management's approval of ILC's termination in 1965, was the extent to which the Apollo deadline shaped a series of collaborative interdependencies between individuals as well as institutions. When ILC found itself fired by Hamilton in 1965, its first step was to appeal to allies within NASA, such as Crew Systems Division Director Richard Johnston,[18] with whom ILC had built a relationship throughout the young decade.

What ILC learned from Johnston and others was that, while NASA agreed that ILC did not seem capable of producing a successful suit, it was not entirely convinced that Hamilton Standard could either. Instead of switching immediately to Hamilton for the "Block II" Apollo Suit (David Clark suits were being used for "Block I" missions), NASA immediately arranged a performance-based competition between Hamilton and a new David Clark effort, in which both suits would be tested against the standards that Hamilton had helped develop. Hamilton joined with B. F. Goodrich to produce parts of its suit proposal. It was clear to all parties initially that ILC would not be involved in the competition.[19]

On hearing the news, Lenny Sheperd raced to Houston. With him went ILC's new president, Dr. Nisson Finkelstein (who had been hired in part for his own technical background).[20] The men offered NASA a new suit prototype, at the firm's own expense, if the agency agreed to test it alongside the two official contestants. NASA consented. They had six weeks.

THE COMPETITION

For the ILC team headed by Sheperd, the six weeks of competition were the crucible out of which a vastly revised suit design was forged. A new shoulder design borrowed from previous B. F. Goodrich work, as well as ILC research. A completely new method of restraining the expanding pressure bladder to allow easy movement in the hips and thighs was invented and implemented by George Durney. Furthermore, since ILC was not being paid by NASA to assemble the prototype, it was free of some of the most onerous of NASA's own suit design specifications, and owned its own new design innovations outright. When NASA patented the lunar spacesuit several years later, a separate patent was issued to ILC and George Durney for the thigh restraint conceived for the competition effort. Instead of having to create a large helmet as specified by NASA, ILC was able to cut size and weight by taking the head measurements of the most big-headed Apollo astronaut (as it happened, Al Shepard), and increasing them by a half-inch in every direction, building on NASA's own "bubble" helmet research.

After Hamilton Standard's staff left ILC's facilities in March 1965, ILC's Government and Industrial Products Division numbered fewer than 120 people. Of these, a bare dozen could be spared to work on the Block II competition. Thus began the 24-hour shifts and the picked supply room locks in search of the materials needed for the prototype suit—to which NASA would give the designation AX5L. With little time to spare, the engineers abandoned more complex testing procedures for ILC's original technique of producing harlequin-like suits in which alternative arms, legs, and joints were tested side by side in the body of the same garment.

In the preparation of such prototypes, as well as ILC's final competition suit, credit is due not only to the supervising staff—whether trained engineers Mel Case or "hard-knockers" like Durney and Sheperd—but also the particularly skilled female garment workers enlisted into the manic competition process as technicians (see Layer 14). With skills in sewing as well as the complex glueing, molding, and latex assembly processes that comprised the final suit, technicians like Ethel Collins, Evelyn Everett, Beverly Killen, and Delema Austin literally fashioned the team's final product out of every material available to hand.

"EVALUATION AND COMPARISON OF THREE SPACE SUIT ASSEMBLIES"

Testing of the suits from the three companies took place in Houston from July 1, 1965. The details of the testing process, in which the Hamilton, David Clark, and ILC proposals were subjected to 22 separate testing procedures, are exhaustively recorded in NASA Technical Note TN D-3482, "Evaluation and Comparison of Three Space Suit Assemblies."[21] The ILC, David Clark, and Hamilton entries were identified as Suit A, Suit B, and Suit C, respectively. As the eight-sentence conclusion to the report stated: "Suit A placed first."[22] ILC won.

The rest of the 134-page document recounts a litany of interrelated performance variables such as size, weight, mobility, visibility, and temperature, vividly portraying the way in which the suit assemblies were shaped rigorously from interior and exterior. From the suit's inside, human parameters such as cones of vision, mobility and exhaustion determined success or failure (one earlier criterion absent from the 1965 test, "level of bruising," shows both the extent of improvement in pressure suit design, and the punishing toll of earlier efforts). From the exterior, suits were scored exhaustively on their interface, both mechanical and spatial, with the fixed variables of Apollo's design. Scores for stowable volume, shoulder width, and intravehicular operation underscored the extent to which the lunar suit was an adaptation not just to the human body and the lunar environment, but to the limited means by which Apollo's goal was being accomplished.

PLAYING THE SUIT

As exhaustive as the resulting technical note is on the subject of the suit's testing and procedures, its precise language masks the tension, emotion, and social realities of the make-or-break contest.

When informed that Michael Collins would be the primary astronaut involved in the competition tests, ILC deliberately constructed its entry to Collins's dimensions—only to be dismayed as they realized that the astronaut's own punishing schedule of simulation and training would leave him only a few days for engagement with the suit selection process. When the zipper on the internal pressure bladder failed in ILC's entry, the firm had only 12 nerve-racking hours to fly the suit to Delaware on a chartered plane for repairs.

While the extent of ILC's numerical victory in the completion analysis makes clear the very real advantages of their entry—"Suit A" bested its rivals in 12 of the 22 tests, and came last only four times—ILC engineers acknowledged their status as a well-loved underdog helped in the testing process. Hamilton Standard provided their own suit subject for the tests, but ILC was fortunate in that "Jackie" Mays, one of the most skilled suit testers at the Houston facility, was not only similar to Michael Collins in physique, but had a warm relationship with Bill McClane, one of ILC's Houston-based engineers, who was the firm's representative during the weeks of testing. Mays was renowned for his ability to "play" a suit—push it to its very limits of performance without breaking a sweat. On the most critical day of testing in 1965, Mays whispered to McClane: "Tighten up this suit, Bill, cause whatever that H.S. bastard does I'm going to do better."[23]

And it was clear that the ILC suit was much better. Even the restrained language of the final technical report reveals stark differences between the suits' abilities to negotiate Apollo's demands. Describing their performance inside the restrictive LEM interior, the report explains: "In the pressurized suit A, the subject was able to reach all areas inside the LEM … he could reach every part of the ascent engine and tunnel area [by] mounting the ascent engine cover, with considerable ease. … In both of the other suits, he was unable to mount the engine cover. … On the contrary, he was forced to 'thrash about' quite strenuously… reaching about in a rather marginal manner."[24]

The process also involved a series of dramas that did not make it into the public record. When a test subject (in this case also Jack Mays) attempted to climb onto the engine cover inside the Lunar Module cockpit (a necessary maneuver if the side hatch was blocked), the increase in pressure inside the David Clark suit blew off its helmet; a stern appraisal in a subsequent internal memo declared: "NASA is concerned with the safety of *all* test subjects."[25] In the Hamilton Standard Suit, a test subject repeatedly failed to enter the Lunar Module at all, the wide shoulders of the suit stranding the imaginary astronaut on the surface of the moon forever.

While the resulting technical analysis took care to rank all three suits in sequence, reaction in NASA to the ILC suit was unequivocal. "The ILC Suit is in first place," Crew Systems official Matt Radnofsky reported to Samuel C. Phillips, the Air Force General brought in to head the Apollo effort in 1964; "there is no second place."[26]

TEAMWORK AND TRANSPLANTS

When ILC lost the first Apollo suit competition in 1962, they were judged as having the best suit for an astronaut to wear, but being ill-suited organizationally for the management demands of the moon race.

In 1965, ILC was judged again as having the best suit, this time for Apollo's vehicles as well as its astronauts. This time, it was given a chance to clothe itself in new organizational garments as well; but the transformation would not be without its own challenges. Lenny Sheperd and George Durney remained the engineers in charge of ILC's Apollo project, but an enormous effort was undertaken to appropriately expand the 120-person ILC organization.

13.7
A prototype A7L without its Thermal Micrometeoroid Garment (TMG).

13.8
A prototype A7L used to study the attachment of
the Thermal Micrometeoroid Garment (TMG).

In the following months, ILC hired experienced engineers like Walter Lee, whose decades of quality control and process management with the Navy and Boeing reassured NASA at the same time as it came to ensure rigor in ILC's ballooning facilities. The need for experience in the military-industrial vernacular of systems documentation was only underscored by NASA's first responses to ILC's independently produced paperwork. "Re: Pre-Delivery Acceptance Inspection and Test Plan and Procedure for Apollo EMU Garment CEI Program," one NASA memo opined, "The subject type 1 document has been reviewed by NASA and found totally unacceptable."[27] ILC was in the ironic position of being able to clothe a man for the moon, but not to equip itself for the bureaucracy of the space race itself.

HAMILTON AND HYBRIDS

ILC's AX5L "Suit A" achieved a hard-won and decisive victory over Hamilton Standard's AX5H "Suit C" in 1965. The strain and stress of their unsatisfying three-year contractual relationship further emerges in accounts from those involved in the Apollo suit process. And yet, while ILC adapted far better to the physical body of the astronaut (as well as to the emerging physical limitations of Apollo's strategy) with its 1965 design, the entire package that sustained astronauts on the moon would remain a hybrid. When ILC won the 1965 test, NASA did not cancel Hamilton's suit contract, but simply trimmed it to reflect changing circumstances. Hamilton remained a highly successful supplier of the crucial portable life support system (PLSS) that hung like a backpack on the ILC suit. And Hamilton's own innovations in suit comfort and mobility—notably the water-tube Liquid Cooling Garment—were transferred, via NASA, to ILC's production line.

THE MOON SUIT PLAYS FOOTBALL

In 1967 the Dover, Delaware spacesuit operation finally split from the "Playtex" brand and became a company in its own right. Under a larger corporate umbrella (now the Glen Alden conglomerate), Abram Spanel's original International Latex Corporation was split into the initialed "ILC Dover Inc.," maker of spacesuits, and the new "International Playtex Corporation," which continued bra and girdle sales. Yet, in managing a final threat to their supply of Apollo suits in early 1968, ILC Dover revealed a flair that clearly drew from the firm's marketing origins.

In 1967, a fatal fire brought an early end to Apollo's "Block I" of missions, leaving astronauts Grissom, White, and Chafee dead on the launch pad. In a decision that ran back to the 1964/65 delays in Hamilton's Apollo contract, the three were wearing David Clark suits. While the fire would result in material transformations to all manned Apollo vehicles—including ILC's AL7 suit—the tragedy wrought organizational changes on Apollo as well. As of the next flight, Apollo 7 in October 1968, systems were slated to be "all up" for Block II—with the result that the David Clark Gemini/Apollo suit was never used in space.

The 1965 competition had been explicitly mounted for these "Block II" Apollo missions, which included the first landing and initial explorations. At the time, it was imagined that a further competition would be held for the final "Block III"—which was planned to include long expeditions on a lunar rover and up to two weeks on the surface. Litton Industries and Air Research, two longtime developers of "hard" suits, developed soft and/or stowable versions of their constant-volume designs, which could ingeniously fit inside of Apollo's confines but whose surfaces became rigid upon inflation. It was predicted that one or the other of these, last, hard suits would clothe Apollo's final astronauts. Yet this was not to be. At the same time as such suits were being tested and developed in Houston and elsewhere, larger U.S. budget pressures brought a limitation in Block III's scope, even before the first Block II flight had launched. Eventually, only Apollos 15, 16, and 17 would be launched.

Conscious of the continuing threat to their position as suit contractor, ILC continued to fund its own internal development work just as it had since the 1950s. The result of a multiyear effort at greater mobility and endurance was termed the "Omega" suit by ILC, and became the AL7B used on the moon's surface during Apollos 15, 16, and 17. When ILC presented the Omega to NASA in 1968, however, the crucial document in convincing the Agency to stick with ILC over hard alternatives was not reams of paperwork, but rather a strip of motion picture film.

Seeking to silence skeptics, the ILC Omega team had produced the film, containing no technical data whatsoever. Driving a loaded company station-wagon from their factory, the team set up a 16mm cameraman, test subject, and life support pressure system at the Dover High School football field. For several hours—the length of a late-Apollo EVA—the subject in a pressurized suit eschewed the more technical motion studies of NASA's internal analysis for a full-fledged game of football—running, throwing, catching passes and punting, falling and bouncing to his feet. And as became clear on watching the films, the suited subject's attempts were at the very least equivalent to those of an engineer in shirtsleeves and slacks who joined him on the field. NASA adopted the suit soon thereafter, terming the highly modified suit the AL7 "B" to avoid the need for a further contract bid. ILC Dover, née Playtex, had won the Apollo game.

CONCLUSION

NASA's 1962 decision to create a hybrid of Hamilton Standard and ILC presumed that it would be possible to graft ILC's designers onto the framework of a traditional aerospace contractor. The graft did not thrive, but the resulting adaptation was in many ways more remarkable. With the same ingenuity that allowed it to create an operable lunar suit prototype in 1962, ILC adapted itself to the shifting organizational circumstances that surrounded it, adopting new procedures and learning foreign tongues of testing, evaluation, and organization "control."

In so doing, however, ILC was not so much transforming itself as an organization, but extending the strategy of adaptation and survival that served its 1962 moon suit prototype so well to begin with. Deploying multiple, adaptive layers, ILC's suit weathered at least three separate atmospheres; each in some ways was stranger than the next.

The first atmosphere was that of a lunar vacuum, and against it ILC offered, from the mid-1950s, a unique understanding of a new kind of protection against outer space, one that would allow work, movement, and even the ability to walk on new celestial bodies. Having produced the most successful suit of the many examined in 1962, ILC with Hamilton had to adapt a successful prototype to the emerging specifics of the lunar mission—the physiological realities of the astronaut's own bodies, as well as the emerging physical realities of Apollo's minimal spacecraft. Such a process became quickly weighed down by the labors of collaboration between the two

13.9
Pressurized A7LB in performance demonstration on
Dover High School football field, 1967.

13.10

Apollo 17 astronaut Harrison Schmitt stands beside the flag in what is likely the last photograph of an Apollo astronaut on the Moon. He is wearing ILC's "Omega" suit, termed the A7LB to allow its procurement under the same contract as the earlier A7L.

very different organizations. Yet in 1965 ILC was able to again produce the most successful lunar prototype in just six weeks.

This prototype succeeded, in turn, because of its adaptation to a second kind of atmosphere—the emerging and very real physical reality of Apollo's schedule, equipment, and extreme physical limitations.

Finally, in successfully executing its "prime" suit contract from 1965 to 1970, ILC performed an adaptation that NASA's planners had considered unlikely in 1962—changing both the scale and procedures of its operation to adapt to the organizational atmosphere of Apollo's massive bureaucracy.

The relative ease with which ILC was able to accomplish these successive and remarkable adaptations should provoke the examination not only of what was so remarkably different about the organizational, architectural, and physiological challenges to which ILC adapted, but also of what was in the end so remarkably similar about them.

In 1962, ILC produced an entirely new kind of architecture—an environment, or protective layer, that would interpose itself between earth-adapted man and the complex hostility of the lunar surface, and allow him to walk and work. In 1965, ILC directly borrowed the same adaptive techniques to produce a suit that would meet an even more complex understanding of the lunar surface and simultaneously adapt itself to a brand-new complex architecture—the technical limitations of Apollo. In 1968, as Apollo's ambitions extended to greater motion and travel, so did ILC further adapt its suit and successfully ensure its continued adoption within NASA's complex hallways. Whether extending the limits of human physiology to a lunar vacuum, or adapting the pressure-suited body to the couches and crawlways of Apollo's internal architecture, ILC was able to fashion the astronaut's body with a continuing ingenuity and success that surprised all observers and competitors. In order to do so, ILC had to construct an even more complex, if ephemeral, set of clothes to protect and sustain its own organizational body.

In 1962, ILC's successful team was an improbable dozen of seamstresses and engineers, led by a car mechanic and a television repairman. A much stranger species than Hamilton's Air Force-led "tiger" team, this group was well-fitted to the challenges of lunar spacesuit production, even as its outlines appeared impossible to prospective collaborators and customers at Hamilton and NASA. While an initial attempt by NASA to adapt this strange animal to its own organizational atmosphere backfired, and indeed caused much discomfort, the experience would allow ILC to adapt a new architecture of its own. This adaptation, the self-described "buffer" of new engineers, translators, and facilitators protecting the informal expertise at ILC's core, was in its own way just as essential a garment in Apollo's success as the resulting AL7 itself. Instead of the penetration of organizational systems into the body (as conceived by Clynes and Kline and their heirs) we see precisely the reverse—the extension of messy, open-ended techniques of clothing, layering, and stitching from real fabric into organizational fabric as well.

Notes

1. National Aeronautics and Space Administration History Office, Oral History Interview with Mr. Mel Case, Senior Design Engineer, ILC Industries, Inc. and Mr. Leonard Sheperd, Vice President of Engineering, ILC Industries, Inc., April 4, 1972 (transcript of tape recording), 8.

2. Ibid., 39.

3. Interview with Homer S. Rheim (former CEO, ILC Industries, Inc. and design engineer, Apollo Spacesuit program) by Nicholas de Monchaux, October 23, 2002, tape recording.

4. Telex from NAA S&ID Downey, CA to NASA MSC, Houston, Texas, December 8, 1963, Project Control Files, National Aeronautics and Space Administration, Manned Spacecraft Center, Engineering and Development Directorate, Crew Systems Division, Record Group 255-E171B, National Archives and Records Administration, Southwest Region (Fort Worth, TX).

5. Interview with Bill Ayrey (Company Historian, ILC Dover) and Amanda Young (National Air and Space Museum, Smithsonian Institution) by Nicholas de Monchaux, January 15, 2006, tape recording.

6. Kenneth S. Thomas and Harold J. McMann, *U.S. Spacesuits* (Berlin: Springer, 2006), 97.

7. Robert Gilruth (for George Low), Memorandum EC911BG211 to Dr. George Meuller, NASA Headquarters, "Procurement plan for the Apollo Extravehicular Mobility Unit and EMU ground support equipment development and fabrication," September 20, 1965, Project Control Files, NASA, Crew Systems Division.

8. NASA History Office. Ibid., 20.

9. Rheim interview by author.

10. Ibid.

11. Case and Sheperd interview transcription, 30.

12. M. I. Radnofsky, Memorandum to Record, December 21, 1964, Project Control Files, NASA, Crew Systems Division.

13. Ibid. In the same note, Radnofsky is at pains to point out that the collected NASA officials offered to "give ILC another chance if they wanted," and "made no recommendation or commitment in writing or verbally to [Hamilton Standard] as to our concurrence or nonconcurrence with the decision."

14. Gilruth, Memorandum EC911BG211.

15. Ibid.

16. Interview, John McMullen (Lead Engineer, Apollo, ILC Industries), Thomas Pribanic (Apollo Configuration Management, ILC Industries), and Richard McGahey (Quality Engineering, Apollo, ILC Industries) by Nicholas de Monchaux, May 30, 2006, tape recording.

17. Ibid.

18. Johnston had acquired a reputation as a stickler for fair evaluation since his first days at NASA during the Mercury Program. The Air Force had offered its David Clark X-15 suit as adequate for Mercury and, until Johnston's arrival, the suit was scheduled to be used. Arriving in the young crew systems division, Johnston insisted on a comparative testing of prospective suits. Despite the Air Force's insistence to the contrary, he determined that the X-15 suit was poorly adapted to Mercury's ventilation system, leading to the choice of the B. F. Goodrich Mark-IV-derived suit instead. After the Air Force protested, NASA backed Johnston and the incident set an early precedent for performance over institutional loyalties in the young space program. National Aeronautics and Space Administration, Oral History Project, Interview with Richard S. Johnston, Houston Texas, August 11, 1998 (transcript).

19. "Apollo EMU Procurement Package," March 2, 1965, Project Control Files, NASA, Crew Systems Division.

20. McMullen et al., interview.

21. R. L. Jones, "Evaluation and Comparison of Three Space Suit Assemblies," NASA Technical Note TN D-3482 (Washington, D.C.: National Aeronautics and Space Administration, 1966).

22. Ibid., 92.

23. McMullen et al., interview.

24. Jones, "Evaluation and Comparison of Three Space Suit Assemblies," 24.

25. Memorandum to Apollo Support Office/EC9 from Crew Performance Section EC5, "Response to PB-2-108," June 30, 1965, Project Control Files, NASA, Crew Systems Division.

26. Case and Sheperd interview transcription, 32.

27. NASA—which knew by this point that ILC was the only company able to build a successful Apollo suit—adopted a schoolteacher's tone in the rest of the directive: "ILC is requested to send appropriate representatives to NASA to participate in rewrite of the document with NASA assistance. These representatives are required as soon as is possible and should be prepared to stay until rewrite is complete." Memo to L. F. Shepard, International Latex Corporation, from Richard S. Johnston, Chief, Crew Systems Division, June 22 1966, Project Control Files, NASA, Crew Systems Division.

Layer **14** HANDMADE

14.1
Jane Butchin, Delema Domegys, Inspector Mary Todd,
and others on the shop floor at the Dover, Delaware,
ILC plant, June 28, 1967.

When she arrived at ILC's spacesuit plant in 1965 or 1966, likely transferred from making bras, girdles, or diaper covers for Playtex,[1] a new seamstress would be greeted by her shop-floor supervisor, and "taught to sew again from scratch."[2] While sewing a spacesuit was in many ways similar to sewing a bra or girdle—the Singer machines and pattern templates were identical—many details of the process were, by comparison, out of this world. A garment assembly line manufactures to a tight tolerance, and a couture house even tighter. The tolerance prescribed for ILC's suit assembly line, however, was derived from NASA-mandated systems engineering guidelines, and was much stricter than any normally brought to bear on a handcraft. NASA's standards pushed the very limits of the equipment used by ILC, as well as the very limits of the seamstress's own technique. The tolerances allowed—less than a sixty-fourth of an inch in only one direction from the seam—meant that yard after yard of fabric was sewn to an accuracy smaller than the sewing needle's eye. To achieve such precision, many women used a modified treadle that, instead of starting and stopping the Singer's operation, fired one stitch per footfall through the multiple layers of a suit's surface. For the hundreds of feet of seams in each suit, this meant venturing stitch by tiny stitch across the length of a football field, with a single misstep leading to a discarded suit.

At the same time ILC's seamstresses were being asked to meet unprecedented precision, they were denied the traditional tools used to maintain sewing accuracy—pins and other temporary fasteners. To a garment whose reliability depended on an impermeable rubber bladder, mechanical aids like pins were an inherently risky proposition. In 1967, after a single pin was discovered between the layers of a suit prototype, an X-ray machine was installed on the shop floor to regularly scan the suits for errant fasteners. Especially thereafter, the use of pins was highly discouraged and, where allowed, highly regulated. The most valued seamstresses were those like Roberta Pilkenton, who could sew together the outer Thermal Micrometeoroid Garment's (TMG) 17 concentric layers, with hundreds of yards of seams, without a single tool except her own guiding fingers. Those who persisted with pins as an aid to assembly were required to check out a numbered set from a supervisor at the beginning and end of their shift, accounted for daily. Those flouting regulations and bringing extra pins from home could, notoriously, find one of them pricked into their backside by an irate supervisor.

If the precision involved in sewing concentric suit layers seems fastidious, then the task facing the "gluers"—who assembled each layer of the suit's concentric surfaces together before final sewing—was even more finicky. Layering flexible latex, and whisper-thin layers of Mylar, Dacron, and Kapton, they used paintbrushes and specially formulated glue to assemble diaphanous sheets into man-shaped assemblies. The tolerance for assembly was no more than the thickness of a single Mylar layer. Each of the 16 glue-assembled layers needed not just to fit the astronaut's body shape, but, like dolls in a Russian *matrushka*, to be infinitesimally larger than the layer it contained.

Any visitor to ILC's Dover plant—or the new Frederica, Delaware facility into which operations expanded in 1966—could observe the care and craftsmanship of those who glued and sewed the A7L's layers together. What they could not freely observe, however, was the most closely guarded craft of ILC's suit assembly process, the "dipping" of layers of rubber to form the ribbed sections, or "convolutes"; it was these assemblies that gave ILC's suit its essential mobility. The "dippers" who accomplished this task were, like their colleagues behind Singer machines, using skills found

14.2

An unknown ILC employee fabricates the shell to a backpack assembly.

14.3

Big Moe sewing machine. The seamstress working it, Hazel Fellows, an ILC group leader, assembles the shell, liner, and insulation of a Thermal Micrometeoroid Garment cover layer.

elsewhere in the Playtex organization. Until 1966, pipes of liquid latex ran to the dipping room from the same tanks supplying girdle and bra assembly lines. As with the rest of the spacesuit makers, however, their skills were brought to an extraordinary level of craft and precision.

The elaborate process of assembling a convolute started with a collapsible formwork around which layers of natural rubber were formed through repeated baths in a latex-bromine mixture. Into this layered surface were imbedded the caterpillaresque restraining rings, which allowed the convolute to retain its shape along the arm or leg's axis when several pounds of pressure inflated the suit against a vacuum. Imbedded above and below the restraining rings was a thin layer of nylon tricot—the same cloth that formed Playtex's brassieres. The sheer fabric, imbedded in the rubber, allowed the convolute to be flexible while pressurized, without ballooning in volume. This delicate assembly had been the key to ILC's suit designs since the early 1960s, and its equally delicate manufacture was the company's most closely guarded industrial secret. Even so, it was more a craft than a science, and only three or four employees had the right "touch" to consistently produce usable components.

In retrospect, however, ILC's unique skill seems to have gone beyond these individual crafts, and into the delicate art of their collective synthesis. Crucial to this larger success seems to have been the professional respect accorded to, and practical collaboration engaged in with, ILC's crafts-women. Indeed, some of ILC's most effective engineers, such as Robert Wise, took weeks of sewing lessons from the seamstresses to better understand how fingers, fabric, and thread interacted to build up the suits' complex assemblies. In practice the craftswomen were allowed, and even encour-aged, to suggest improvements in procedures and assemblies as they were continually developed.[3]

Emblematic of this culture were the late-night collaborations between seamstress Eleanor Foraker (pulled from Playtex's diaper cover assembly line in 1964) and Leonard (Lenny) Sheperd, project head of Apollo suit development for ILC. Striving to meet deadline after deadline, Foraker would often be sewing late into the night at the Dover plant. In the final stages of a suit's manu-facture, the multilayered, man-shaped assembly could not be folded or squashed under a normal sewing machine, and instead had to be sewed on one of two Singer machines—dubbed "Big Moe" and "Sweet Sue"—modified to have an elongated arm and massive sewing bed so an entire suit could be moved under its needle. For each of these early deadlines, Sheperd himself would stay up with Foraker, helping to slide and rotate the suit during the final stages of assembly, and quiz-zing the seamstress on her technique and expertise.[4] (The moment might be akin to Simon Ramo or Dean Wooldridge of TRW assisting in the soldering of a guidance system circuit board.)

And yet, even as ILC managed to integrate technique and technology within its own corporate body, it proved enduringly incapable of fully adapting to the organizational atmosphere of Apollo. The most prominent failures surrounded the procedures of systems management, pioneered by Ramo and Wooldridge with Bernard Schriever in the ICBM programs of the 1950s and carried, with ICBM hardware, directly to NASA.[5] From the time of its June 1965 victory over Hamilton Standard, to well into the first flights of Apollo spacecraft in 1968, these problems taxed the patience of NASA managers even as they led to a series of important shop-floor adaptations on the part of ILC.

14.4

Velma Breeding installing a bladder into the
PGA restraint boot.

14.5

Arlene Thalen, an ILC inspector, examining Mylar
insulation layers. The layers were glued from
patterns, and then inserted one inside the other
to form the insulation layers of the Thermal
Micrometeoroid Garment cover layer.

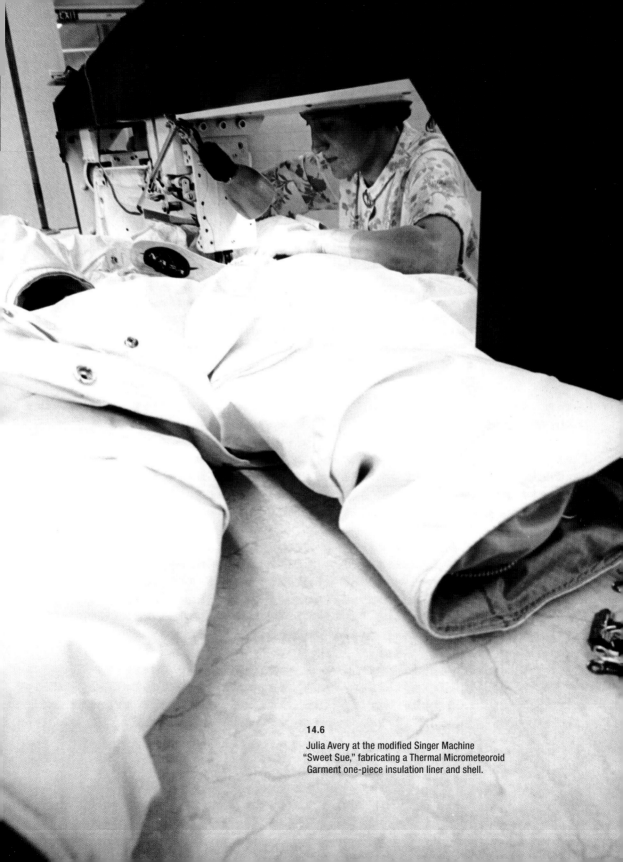

14.6
Julia Avery at the modified Singer Machine
"Sweet Sue," fabricating a Thermal Micrometeoroid
Garment one-piece insulation liner and shell.

SEWING PAPER

In 1964, a procurement official based in the lunar lander effort complained to NASA headquarters of Apollo suit prototype deliveries: "The subject shipments have been arriving unaccompanied by proper shipping documents. … After 1 August 1964, all such material not properly shipped will be subject to rejection and will be considered as not having been furnished at all."[6] The paper version of the spacesuit rivaled, and even supplanted, the physical object—at least in institutional terms.

Particularly vexing to NASA, then, was ILC's avoidance of paperwork—not just in the everyday transport of spacesuits, but in their manufacture. From the first days of its spacesuit prototypes, ILC had eschewed traditional engineering media—measured drawings, charts, and memoranda—in favor of those derived from its soft-goods heritage: patterns and instructions. While ILC competitor and collaborator Hamilton Standard insisted on the same drawing-based procedures in assembling "soft" components as in the rest of its "hard" aerospace business, ILC's attitude came through in a seamstress's comment to a NASA technical team in 1968: "It might look all right on that piece of paper, but I'm not going to sew that piece of paper."[7]

Yet it was reams of paper that fed the maw of the systems engineering machine. ILC's new managers worked hard both to convince NASA of the firm's established methods and to pack additional, protective layers of bureaucracy around existing procedures, all without compromising the fragile craft of its own expertise. ILC's initial victory in the paper tug-of-war was to convince NASA's Crew Systems Division (which governed the design of the Command Module and Lunar Excursion Module (LEM) as well as spacesuits) to modify its requirement for constantly updated drawings as the day-to-day documentation of the design process. ILC successfully proposed a more formalized version of its existing procedures, recording both patterns and assembly procedures on a regular schedule.

The physical patterns that guided the cutting of the countless cloth layers in the ILC suit were thin layers of manila card stock, constantly subject to change as the fit between pieces and layers was finessed. Starting in 1965, however, the latest copy of the cardboard outlines would be regularly flown to a lab in Philadelphia, where they were photographed in silhouette against a light table, then reduced to microfilm to form a permanent record of the suit's shifting outlines.[8]

At the same time, ILC's procedural notes on sewing became formalized as so-called Tables of Operations (TOs). These were written step by step just like a cookbook recipe or model kit, recording every stage in the suit's complex assembly. Seamstresses—who often understood how fabric behaved better than the engineers writing the instructions—had to petition for a change in the TOs before they were allowed to change the craft of manufacturing, allowing ILC and NASA to keep track of their otherwise mysterious motions.

Despite such advances, NASA's Crew Systems Division (CSD) was as regularly critical of ILC's systems management procedures as they were effusive about the suits ILC produced. In 1966, ILC was issued the first of several official rebukes from CSD director Richard S. Johnston: "It has come to the attention of Crew Systems Division that your failure reporting and analysis systems has [sic] not been meeting minimum expectations of quality and punctuality."[9]

A much larger drama played out over the complex issue of "configuration management," a system developed in the Atlas Missile program to precisely track the origin and details of each part in a complex technological artifact. The procedures were an enormous challenge for ILC. It was hard enough in a North American Aviation or Grumman factory to keep track of the origin, assembly, and destination of every spacecraft part; NASA demanded the same paper trail be in evidence for the 4,000-plus pieces of fabric in the 21-layer Apollo suit as well. Such a requirement—a process that could treat every thread, nylon tape, and fabric swatch in the suit with the same administrative tenacity as individual aluminum assemblies—proved difficult to meet.

In February of 1967, with the delivery of the first three A7L suits, matters came to a head. NASA initially refused to accept the suits, not for any physical defect in their manufacture, but rather because their configuration—the position, size, and origin of every part—was not adequately documented. A weeklong crash effort employed seven engineers to improvise an acceptable paper trail for the suits, and led to a sharp letter of complaint from Johnston to ILC's new president, Nisson Finkelstein.[10]

A seven-page technical note followed to ILC's Apollo director, Lenny Sheperd. The memo announced: "It has been determined that requirements and intent of Apollo Configuration Management Manual, NPC 500-1, and ANA Bulletin 445 are not being met. Investigation into the causes of noncompliance indicates a lack of authority or clear understanding of the methods, processes and controls required of the contractor personnel [sic] for proper implementation." Following the finding of noncompliance, a list of separate action items was issued. These included the implementation of 14 new procedures, and further demanded that drawings and other traditional engineering media be introduced into ILC's procedures.[11]

It was to adapt to this context that ILC added to its roster of sewers, dippers, and gluers a new legion of employees trained in systems engineering culture. As we have seen, these individuals talked among themselves of being an essential "buffer" between the systems culture of Apollo and the culture of craft at ILC.[12] In the heat of the space race, however, ILC could not hire as many systems engineers as it needed, and so devised a plan for a literal transplant of skills.

While traveling back and forth to Texas, Lenny Sheperd had come to know key managers at Ling-Temco-Vought (LTV), at the time a massive aerospace conglomerate (and one of the original eight bidders for the Apollo suit).[13] Admitting "we knew all there was to know about suits, but we sure didn't know very much about the paper,"[14] Sheperd arranged to contract 56 LTV specification writers, configuration managers, and systems engineers to come to Dover. Once the LTV managers arrived in Delaware, Sheperd arranged for the "smartest" new ILC employees to shadow each one to the point at which the skills were imbedded in his own people. Under this arrangement, the first redundant LTV manager was sent back to Houston less than three weeks after his arrival (the last to leave, manager Gary Raper, was contracted for 14 years).[15]

Working together, the ILC and LTV engineers attempted several methods to adapt ILC's soft-suit process to NASA's military-industrial vernacular. The first of these used a Lockheed Martin-designed framework for systems engineering, and proved unworkable when a foot-high stack of onionskin teletype paper was repeatedly printed for each suit. The final system deployed a series of computer punch cards, handled along with the suit, hand-punched using a predetermined system of codes for parts, suppliers, and assemblies. Only if NASA demanded a full enumeration of a single suit's assembly—as in the case of a mission failure or testing problem—were these punched cards digitally processed to produce a full stack of paperwork. While such a step was rarely necessary, the system produced a working relationship with NASA; the perceived capacity to produce a mountain of paper for each suit allowed the physical production of forests of forms to be avoided.

Even by 1968, however, NASA's attitude toward ILC appeared unchanged. A May 1968 NASA report remarked of ILC's efforts that they were "About the same as reported for previous periods … [an] improvement in product quality, with the performance of engineering and configuration management groups … marginally satisfactory."

Yet in the same note there are signs of growing détente. While ILC is criticized for a lack of "discipline," for the first time it is allowed that the problems of producing suits to systems engineering standards result "in part by the nature of the contract end items under production."[16]

With such a philosophical shift—the recognition that the A7L was fundamentally different from other artifacts of the space race—came procedural accommodations as well. The result was a complex compromise between the intimate realities of spacesuit design and the institutional regulations of the space age.

ILC, for its own part, increasingly produced documents to NASA specifications. None were more important than the Acceptance Data Pack (ADP) that accompanied each suit. A raft of collated forms, specific to each suit, the ADP was literally incorporated into the physical case surrounding the suit on delivery, slotting into its cover. An institutional shield, the ADP tailored the spacesuit to its institutional client, providing reams of requested data. As a result, in the words of one engineer, it was "more real" to NASA's systems engineers than the physical suit it attempted to describe.

If ILC slowly suited itself to NASA's norms, then the systems engineers also, gradually, accepted some of ILC's departures from military-industrial norms. After an extended on-site meeting in the summer of 1968, a crucial decision was taken to allow the artifacts of handcraft—the patterns and procedures of suit construction, or Tables of Operations—to temporarily substitute for more traditional engineering drawings. After suit development and manufacture, ILC would not be required to issue separate drawings solely to satisfy NASA. Such drawings would still be produced to satisfy the contractual requirements of ILC, but as beautifully precise forensic descriptions—not instructions.[17]

These elaborate drawings of hand-crafted clothing would have their own particular life as objects of handicraft—in a series of lithograph collages by Robert Rauschenberg, collectively entitled *Stoned Moon* for the sense of delirium surrounding the event, as well as the more literal lithographic stones into which the image was worked.

Invited by NASA administrator James Webb to attend the launch of Apollo 11 in July 1969 as part of the official NASA Art Program (which had previously focused on more "realist" painters like Norman Rockwell), Rauschenberg requisitioned a range of visual material from NASA, in which the ILC drawings were included. In a fever of activity in late 1969 and early 1970, Rauschenberg produced 34 lithographs in 34 days at the Gemini GEL Printworks in Los Angeles.

When Rauschenberg had first been encouraged to use lithography in the early 1960s, he had disparagingly disagreed; the second half of the twentieth century was "no time to start writing on rocks."[18] In the end he became one of the medium's most noted artists, applauding the "receptive, organic, flexible" quality of the technique.[19]

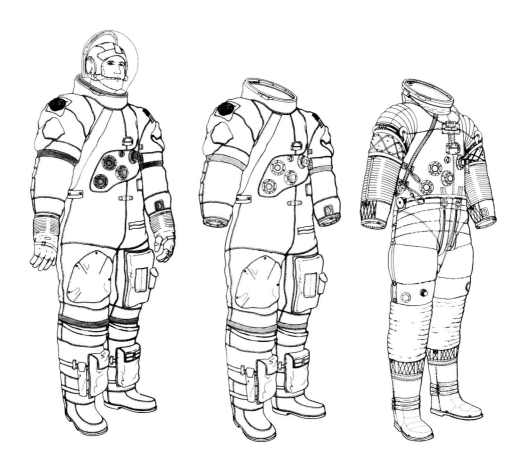

14.7

A composite of the final delivery drawings produced
by ILC from 1970, showing an A7LB suit as worn
by an astronaut with gloves and helmet (left),
with Thermal Micrometeoroid Garment (TMG) layer
(middle), and showing only Pressure Garment
Assembly (PGA) (right).

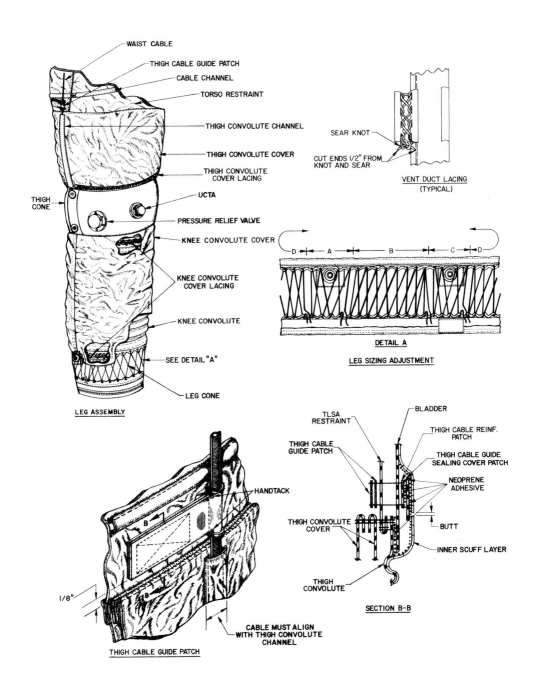

WAIST CABLE
THIGH CABLE GUIDE PATCH
CABLE CHANNEL
TORSO RESTRAINT
THIGH CONVOLUTE CHANNEL
THIGH CONVOLUTE COVER
THIGH CONVOLUTE COVER LACING
THIGH CONE
UCTA
PRESSURE RELIEF VALVE
KNEE CONVOLUTE COVER
KNEE CONVOLUTE COVER LACING
KNEE CONVOLUTE
SEE DETAIL "A"
LEG CONE

LEG ASSEMBLY

SEAR KNOT
CUT ENDS 1/2" FROM KNOT AND SEAR

VENT DUCT LACING
(TYPICAL)

D — A — B — C — D

DETAIL A

LEG SIZING ADJUSTMENT

HANDTACK

1/8"

CABLE MUST ALIGN WITH THIGH CONVOLUTE CHANNEL

THIGH CABLE GUIDE PATCH

TLSA RESTRAINT
THIGH CABLE GUIDE PATCH
BLADDER
THIGH CABLE REINF. PATCH
THIGH CABLE GUIDE SEALING COVER PATCH
NEOPRENE ADHESIVE
THIGH CONVOLUTE COVER
BUTT
INNER SCUFF LAYER
THIGH CONVOLUTE

SECTION B-B

14.8

An illustration of the complex mechanisms for
adjustment present in the PGA leg assembly.

As suited the epic material of their content, the 34 lithographs of *Stoned Moon* pushed the limits of lithographic technology; three of the stones represented the largest prints ever made by the process, taking more than two pounds of ink for each impression.[20] Recalling the visual architecture of Mission Control (see Layer 18), John Cage would liken the hand-crafted lithographs' impact to watching "many television sets working simultaneously all tuned in differently."[21]

While Rauschenberg would use the archetypical image of Aldrin on the moon (figure 1.1 of this book) in a later photocollage, *Signs* (which otherwise commemorated the deaths of Robert and John Kennedy, Martin Luther King, and Janis Joplin), the only face-on image of an Apollo astronaut in the *Stoned Moon* series is a linework illustration of the suit's outer Thermal Micrometeoroid Garment from ILC's suit drawings.

In the course of producing the *Stoned Moon* lithographs at the Gemini Printworks, Rauschenberg invented a new set of techniques to transfer photographs and line drawings to a photosensitized lithographic stone, which he overlaid with a series of hand-drawn lines with a lithographic crayon. In the print entitled *Trust Zone*, Rauschenberg juxtaposes a blackline map of Cape Canaveral's combination of natural and artificial geography with an ILC drawing of the A7L in sky blue. Overlaid atop both line drawings is a December 1903 photograph of Orville Wright making the first powered flight at Kitty Hawk, North Carolina, sixty-six years earlier (his brother, Wilbur, stands to the left of the frame, having just assisted the flyer's path down a railed runway). While the body of this first successful airplane was shipped back to Dayton, Ohio, the local postmaster, William Tate, gathered the remnants of earlier attempts and gave them to his wife. She used the wing fabric—"French Sateen … cut on the bias"[22]—to hand-sew two dresses for their daughters.[23]

S, M, L, XL

This brings us to the most intimate realm of compromise and design in each Apollo spacesuit: the couture-like custom-fitting to a single astronaut's body. To expedite the physical measurement involved, ILC hired Richard Ellis, a former woodworker, to fly around the country and take the measure of each astronaut, so that an individual suit pattern could be developed to suit their body. In the hubbub of the astronaut's training schedules, time for the elaborate series of measurements was not easy to come by—Neil Armstrong met with Harris in a California hotel bathroom.

Subsequent alterations, including even the tightening or loosening of laces and straps, were accomplished in one-on-one "fit-checks" with each astronaut, with forms recording each level of finesse entering into the suit's growing paper trail. Indeed, like any couture customer, the astronauts would often change their minds about fit details, sometimes causing their entire suit to be taken down to its components and carefully reassembled.[24] A final move toward comfort, however, erred more on the side of bras than battlefields. After complaints about comfort, a layer of fuzzy girdle liner was sewn into each garment's pressure bladder.[25]

The custom-fitting of each suit to each astronaut provides a final, illuminating story of the conflicts between a systems engineering culture and ILC's intimate expertise. In every other part of the Apollo effort, the same documentation that charted the 4,000 individual elements of the A7L suit would assign new serial numbers and corresponding paperwork to any adjusted part, or any deviation from the standards set by a part's previous assembly. To NASA system engineers,

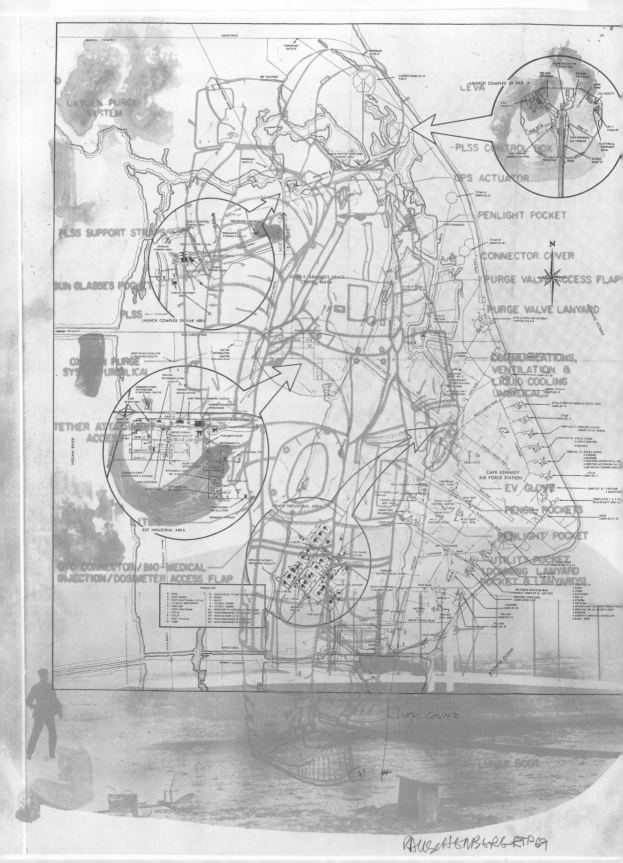

RAUSCHENBERG RTP 69

14.9

Robert Rauschenberg, *Trust Zone* (1969), from
the *Stoned Moon* series (1969–1970). Lithograph,
printed in three colors from two photosensitive
stones and one aluminum plate, 40 x 32 inches;
edition of 65.

Title: "PGA measurements for N. Armstrong"

2 OCT 67 ILC

MEDIUM LONG

Subject	NEIL ARMSTRONG	Location	MSC
Date	10/3/67	ILC Tech.	R. Ellis

Measurement Location	CM.	IN.	Measurement Location	CM.	IN.
Weight	173	LBS.	Upper Thigh Circumference		23 3/4
Height		70 3/8	Mid Thigh Circumference		21.0
Cervical Height		60 1/2	Lower Thigh Circumference		15 1/2
Mid Shoulder Height Right		59 7/8	Knee Circumference		15 1/8
Mid Shoulder Height Left		59 7/8	Calf Circumference		15
Shoulder Height Right		57 3/4	Lower Leg		9 1/8
Shoulder Height Left		57 3/4	Ankle Bone		10 1/4
Suprasternale Height		57 1/2	Scye Circumference Right		18 1/4
Nipple Height		51 3/4	Scye Circumference Left		18 1/2
Waist Height (Back)		43 1/4	Axillary Arm Circumference		12 7/8
Trochanteric Height		36 3/8	Biceps Flexed Circum.		13
Knee Cap		21 1/8	Elbow " "		12 3/8
Center Knee-Floor Height		20	Forearm Flexed Circum.		11
Crotch Height		33 1/2	Wrist Circumference		7
Shoulder-Elbow Length		14 1/8	Sleeve Inseam Right		19
Inter scye Breadth		14 3/4	Shoulder-Elbow Pivot		12 7/8
Biacromial Breadth		16 1/2	Elbow Pivot-Wrist		11 7/8
Shoulder Breadth		19 3/8	Wrist for Finger Tip		7 1/2
Chest Breadth		15	Vert Trunk Circ. Right		68
Waist Breadth		12 1/2	Waist, Front Length		15 1/2
Hip Breadth		13 3/8	Anterior Neck Length		4 1/2
Vert Trunk Dia. Right		25 7/8	Posterior Neck Length		4
Vert Trunk Dia. Left		25 1/2	Waist Back Length		18 5/8
Head Circumference		22 5/8	Gluteal Arc Length		10 1/2
Neck Circumference		15 1/2	Crotch Length		30
Shoulder Circumference		47	Span		71 5/8
Chest at scye		39 1/2	Span Free		—
Chest at nipple		38 1/2	Metacarple 2		8 1/2
Waist Circumference		34 1/2	Extended Arm Length	LEFT 76.7 RIGHT 79.4	
Buttock Circumference		39	Mid-Shoulder/Top of Head		10 1/4

FOOT	Right	Left
Length	9 1/2	9
Instep Length	10	9 1/2
Width	B	B

WEARS B SHOE 9 1/2

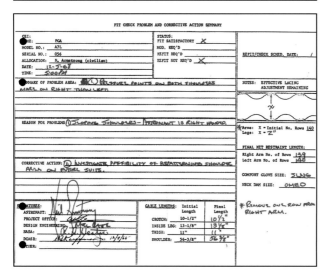

FIT CHECK PROBLEM AND CORRECTIVE ACTION SUMMARY

CEI:
NAME: PGA
MODEL NO.: A7L
SERIAL NO.: 056
ALLOCATION: N. Armstrong (civilian)
DATE: 12-5-68
TIME: 5:00PM

STATUS:
FIT SATISFACTORY X
MOD. REQ'D
REFIT REQ'D
REFIT NOT REQ'D X

REFIT/CHECK SCHED. DATE: /

SUMMARY OF PROBLEM AREA: (1) PRESSURE POINTS ON BOTH SHOULDERS MORE ON RIGHT THAN LEFT.

REASON FOR PROBLEMS: (1) SLOPING SHOULDERS – ASTRONAUT IS RIGHT HANDED.

CORRECTIVE ACTION: (1) INVESTIGATE POSSIBILITY OF REPATTERNING SHOULDER AREA ON FUTURE SUITS.

NOTES: EFFECTIVE LACING ADJUSTMENT REMAINING

x

*Arms: X = Initial No. Rows 140
Legs: X = 2"

FINAL NET RESTRAINT LENGTH:
Right Arm No. of Rows 139
Left Arm No. of Rows 140

COMFORT GLOVE SIZE: SLNG
NECK DAM SIZE: OMEO

*REMOVE ONE ROW FROM RIGHT ARM.

SIGNATURES:
ASTRONAUT: Neil Armstrong
PROJECT OFFICE:
DESIGN ENGINEERING:
NASA:
DCASR: 12/9/68
OTHER:

CABLE LENGTHS:	Initial Length	Final Length
CROTCH:	10-1/2"	10 1/2"
INSIDE LEG:	13-1/8"	13 1/8"
THIGH:	11"	11"
SHOULDER:	56-3/8"	56 3/8"

then, every change to the suit—including both changes in dimension to suit each astronaut, and changes in configuration to enhance individual astronauts' comfort—called for enormous additional documentation. From the ILC perspective, this was tantamount to arguing that a shirt or pair of trousers became a different object when altered, or even buttoned—an argument with its own strict logic, but one that devolved into absurdity the closer the mechanism came to the terminal disorder of the human body.

As with the larger conflicts surrounding ILC's procedures, the resolution was a hard-won compromise. For elements toward the outside of the suit, especially in the covering Thermal Micrometeoroid Garment (TMG), changes followed standard systems engineering protocols. A longer zip or differently shaped piece of Velcro was subject to the same "configuration management" procedures as held sway in any other part of the massive Apollo effort, with up to several changes a day elaborately recorded. Closer to the astronaut's skin, however, the engineering logic evaporated in favor of a humbler logic of clothing.

While each glove finger, anklet, or undergarment was assigned a tracking serial number, it was also given a range of sizes—from small and medium to large—within which its astronaut-specific dimensions could be designated. Once agreed upon, the only problem came with sizing the most intimate part of the suit assembly, the urinary collection device (UCD) that slid over the astronaut's penis. After an "incident" with the first astronaut fitted for the device, the UCD's designations were changed from "Small, Medium, Large" to "Large," "Extra Large," and "Extra-Extra Large."[26]

THE HANDMADE FUTURE

In his seminal 1935 essay "The Work of Art in the Age of Mechanical Reproduction," the German critic Walter Benjamin posits an acceleration and transformation of human existence as a result of revolutionary changes in the technological reproduction of artifacts.[27] Both the elaborate efforts at technological systemization and the ubiquitous images of Apollo's media harvest can be understood as artifacts of the quintessentially modern age that Benjamin heralds. (In their hand-crafted multiplicity, Robert Rauschenberg's collages and combines were one of the century's most cogent artistic commentaries on this seismic transformation in technology and media.)

14.10

The record of one of Richard Ellis's measurement sessions with Neil Armstrong. One complexity in the suit-fitting process was the fact that the astronaut's shifting training regimen often meant changes in his body dimensions and weight.

14.11

A fit check and corrective action summary from Neil Armstrong's A7L spacesuit, the first piece of clothing worn on the surface of the moon.

Yet here is a deep irony—a central, incongruous fact of Apollo's history. For all the systems management efforts spent acclimatizing ILC to NASA's military-industrial milieu, the most heavily reproduced Apollo artifact—the A7L spacesuit seen on stamps, in statuettes, and on screen—is in its essence a throwback, a regression, the sole exception to a rule. For, even as ILC's most dedicated managers would later admit, the forest of paperwork thrown around the suit's craftsmanship was, essentially, "a bit of a smokescreen," hiding a hand-crafted nature. Whether in the "dipping" of convolutes, the glueing of layers, or the sewing of the final suit, ILC's process was so dependent on the individual craftsmanship of its employees that attempts to enumerate precisely the procedures used were inherently impossible. As a seamstress later reflected, "no two people sew alike." While the change of even a sixteenth of an inch in a formalized sewing procedure needed to be debated and recorded in a systems engineering "configuration board," the procedures recorded in paperwork were never precisely those used to assemble the suits. NASA's consistent delight with ILC's suits, and constant frustration with the firm's paperwork, were not contradictory. In fact, they were of a piece. As the many counterexamples to ILC's expertise show, any attempt to fit the human body precisely into the procedures of systems management seemed destined for difficulty. ILC's mastery of the body, and the tucking, tweaking, and tailoring needed for its comfort, placed it inherently at odds with the rest of Apollo's organizational framework. Its success in negotiating that framework—just enough to secure and maintain the suit contract into the 1970s—was, as we have seen, itself a marvel of layered adaptation.

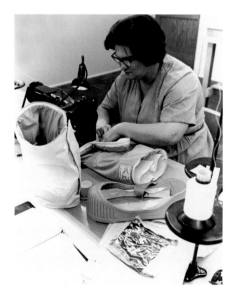

14.12
An ILC seamstress fabricating a TMG outer boot assembly.

Notes

1. Whether through custom or coincidence, no men were employed on ILC's sewing floor.

2. Nicholas de Monchaux, interview with Roberta Pilkenton, ILC Dover, October 27, 2002. Audio recording.

3. Nicholas de Monchaux, interview with Roberta Pilkenton, Bob Ayrey, John McMullen, ILC Dover, October 21, 2002. Audio recording.

4. Nicholas de Monchaux, interview with Eleanor Foraker, ILC Dover, June 2, 2006. Audio recording.

5. For an extended discussion of the importance of Ramo, Wooldridge, and Schriever's techniques for Apollo, see Stephen B. Johnson, *The Secret of Apollo: Systems Management in American and European Space Programs*, New Series in NASA History (Baltimore: Johns Hopkins University Press, 2002).

6. Memo from Samuel Gentile, RABPO, Bethpage, to A. C. Wilder, Procurement & Contracts Branch, MSC Houston, "Re: Shipment of Space Suit Assembly," June 24, 1964, Project Control Files, NASA, Crew Systems Division.

7. Nicholas de Monchaux, interview with Roberta Pilkenton, October 2, 2002.

8. Interview with John McMullen, Thomas Pribanic, and Richard McGahey, Dover, Delaware, June 2, 2006. Audio recording.

9. Memo from Richard S. Johnston to International Latex Corporation Government and Industrial Division, May 23, 1966, Project Control Files, NASA, Crew Systems Division.

10. Letter from Richard S. Johnston to Nisson Finkelstein, March 23, 1967, NASA document EC921IL1849, Project Control Files, NASA, Crew Systems Division.

11. Letter EC911IL1777, March 10, 1967, Memo from Richard S. Johnston to L. F. Sheperd, Project Control Files, NASA, Crew Systems Division.

12. Nicholas de Monchaux, interview with Homer Rheim, Dover, Delaware, October 23, 2002. Audio recording.

13. Ling-Temco-Vought emerged as a contemporary with Litton Industries in the field of large American conglomerates and suffered a similar trajectory. While James Ling, the company's founder, enjoyed a rags-to-riches life story and had by 1969 built a 2.7-billion-dollar business, his firm suffered a similar loss in confidence as Litton, and by 1970 Ling found himself ousted from his position as CEO. See Stanley H. Brown, *Ling: The Rise, Fall, and Return of a Texas Titan* (New York: Atheneum, 1972).

14. "Interview with Mr. Mel Case, Senior Design Engineer, ILC Industries Inc., and Mr. Leonard Sheperd, Vice President of Engineering, ILC Industries" (transcript, 24), NASA Oral History recorded April 4, 1972, courtesy NASA History Office.

15. Interview with McMullen, Pribanic, and McGahey, June 2, 2006.

16. MSC Management Document, Contractor Performance Evaluation (Technical). Contract NAS 9-6100, 5/1/68, Project Control Files, NASA, Crew Systems Division.

17. Interview with McMullen, Pribanic, and McGahey, June 2, 2006.

18. Quoted in Esther Sparks, *Universal Limited Art Editions: A History and Catalogue, the First Twenty-Five Years* (Chicago: Art Institute of Chicago, 1989), 219.

19. Robert Rauschenberg, "Robert Rauschenberg: A Collage Comment," *Studio International* 178, no. 917 (December 1969): 246–247.

20. Jaklyn Babington, *Robert Rauschenberg Stoned Moon Series* (Canberra: National Gallery of Australia, 2010), 7.

21. John Cage, "On Robert Rauschenberg, Artist, and His Work," in Cage, *Silence* (Cambridge, MA: MIT Press, 1970), 98 (first published in *Metro*, Milan, 1961).

22. Lynanne Westcott, *Wind and Sand* (New York: Harry N. Abrams, 1983), 28.

23. Editors of *Time-Life*, *Our American Century: Century of Flight* (Alexandria, VA: Time-Life Books, 1999), 35.

24. Interview with Foraker, June 2, 2006.

25. Interview with Al Gross, ILC Industries (transcript), NASA Oral History Project, recorded April 4, 1972, courtesy of NASA History Office.

26. Interview with Rheim, October 23, 2002.

27. Walter Benjamin et al., *The Work of Art in the Age of Its Technological Reproducibility, and Other Writings on Media* (Cambridge, MA: Belknap Press of Harvard University Press, 2008), 21.

Layer **15**   HARD SUIT 1

"Litton Shoots for the Moon,"[1] proclaimed the April 1958 issue of *Fortune* magazine. Detailing the fortunes of the stratospherically successful company—sales had gone from $3 million to $100 million in just four years—*Fortune*'s feature is lavishly illustrated by paintings of the company's "space laboratory." A full-page spread shows the laboratory's designer, Siegfried Hansen, inspecting "strange gear … the suit worn by the operator."

Unlike the soft suits just being developed for high-altitude flight, and adapted for the space race, the costume modeled by the Litton technician was designed from the outset to be worn in a vacuum. Indeed, the "space laboratory" mentioned in the *Fortune* article was one of the world's largest vacuum chambers. The suit made to operate in this chamber, with its rigid torso and helmet, would form the basis for a series of experimental rigid spacesuits designed by Litton for NASA throughout the 1960s. Like the company itself, the suits were admired inside and outside government, and provided stiff competition to the soft suits developed by ILC.

Even more than their (near) success in the bureaucracy of NASA, however, the Litton suits proved enduringly successful (as in their *Fortune* début) in becoming a popular image of spacewear. Their hard, gleaming surfaces fulfilled a science-fiction fantasy of spaceflight; sleek, efficient, functional, and firm. The vision was both more physically substantial and more stylistically masculine than the layered soft goods of Playtex. And so the failure of Litton's suits, and their descendants, to travel on Apollo spacecraft would be a particular frustration for the engineers and industrial designers who crafted their smooth, interlocking shells.

The story of the Litton suits, however, does not start in outer space. It begins in a quite different kind of inner space: the tiny void of an electronic vacuum tube. Litton's top executives, as well as many of its engineering staff, had split from Hughes Aircraft, whose tube-based avionics for planes and bombs had become central to military aviation in the Second World War. Litton's head, Charles Thornton, had an even longer trajectory in and out of government.

"TEX" THORNTON

Charles "Tex" Thornton had escaped from hardscrabble origins in Texas to success in Washington's booming Depression-era bureaucracy, becoming a successful accountant in the Federal Housing Administration. With the onset of war, Thornton sought employment with the new Assistant Secretary of War for Air, a fellow Texan. "Mr. Secretary," Thornton challenged Robert A. Lovett, "we're putting a tremendous effort into enemy intelligence. But how are we doing at friendly intelligence—reliable information about us?"[2]

The Office of Statistical Control (OSC) that Thornton founded in the Air Corps would become legendary. One of Thornton's first moves was to enter into a strategic partnership with the Harvard Business School (whose usual student body was dissolving into the ranks of the war effort). Recruiting then-Assistant Professor Robert S. McNamara, Thornton transformed the institution into a training ground for thousands of "statistical control officers." Schooled in the latest techniques of information management, the men would form a bureaucracy of surveillance in the Air Forces to report the flow and management of men and materiel; military typewriters were pulled apart and reassembled with yardlong carriages to manipulate the massive "spreadsheet" pages central to Thornton's numerical analysis.

Over the course of the four-year conflict, the OSC's analyses became used not just to track the movement of men and material, but to shape supply chains and time surprise attacks.[3] Not just studies after the fact, they became a central tool in the conflict. After the war, Thornton resigned his commission (he had been promoted to colonel for the impact of his information management efforts on the war) and brokered the transfer of his team to the Ford Motor Company.

First called "quiz kids" for their youth, and habit of asking extensive, statistically oriented questions, they later became known as "whiz kids" for the efficiencies they realized in the carmaker's organization. Before their arrival, stacks of invoices were crudely measured with a ruler to estimate Ford's outstanding payments.[4] After their interventions, complex systems of reporting created a moment-by-moment picture of the firm's operation, bringing higher profit margins and more streamlined product delivery.

Thornton's strong personality, however, grated with polite Ford culture, and it was his protégé, McNamara, who became Ford's CEO in 1960. Thornton, instead, took a position with Hughes Aircraft, where, with his new colleagues Simon Ramo and Dean Wooldridge, he became as optimistic about the future of electronics in the military-industrial market as he was pessimistic about his long-term possibilities under Howard Hughes's erratic leadership.

Ramo and Wooldridge would leave to found their own company, focusing initially on ICBM work (see Layer 5). Thornton, by contrast, borrowed money to buy out an existing vacuum tube manufacturer in Los Angeles, Charles Litton Industries. (Initially seeking to change the firm's name to Electric Dynamics, he went back to "Litton Industries" after objections from the giant General Dynamics.) Soon, Thornton's Litton would be one of the most admired companies in America; as underscored by *Fortune*'s 1958 coverage, no part of its operation was more glamorous than its space-related research, a direct descendant of the firm's vacuum tube origins.

"SPACE TELEGRAPHY"

In 1896, the Briton John Ambrose patented the first of the devices that were to become known as "vacuum tubes." Ambrose's invention was a light-bulb-like device, which, instead of containing a filament, held two separate electrodes. Thanks to the chemical nature of the electrodes in the tube, and the vacuum surrounding it, the tube would conduct electricity, but in only one direction. The tube could therefore transform AC into DC current, and was termed the "oscillation valve."

Later, after the oscillation valve had become known as a vacuum *diode* (for its two electrodes), an American registered the first major improvement on the device, the *triode*. In a patent titled "Space Telegraphy,"[5] Lee de Forest showed how, by adding a third, charged electrode to Ambrose's device (resembling a tiny flyswatter between the two existing terminals), it was possible to greatly amplify the dynamics of whatever weak signal was passed through the tube. Such amplification made broadcast radio possible, as the whispering vapor of electronic transmission could for the first time be reliably resurrected into legible sound. Not only could the new tube amplify and control any analog signal, it could, through its controllable grid, be used to record and alter the presence or absence of electrical charge. With their flow controlled by the new electric valves, these tiny electrical ghosts would become the ones and zeroes of digital computing.

While these early diodes and triodes would be replaced in the 1960s by crystal transistors and silicon wafers, it was the delicate glass of the vacuum tube that lay at the heart of 1950s electronics. Which presented a conundrum. On the one hand, seemingly instantaneous electronic processing opened up an enormous new space for scientists, engineers, and technicians, with the thin surface of the vacuum tube the medium through which a new world of information could be processed and observed (in the case of the giant cathode ray tube of a television, literally so). On the other hand, as an executive of a large radio firm explained in 1930, "tube manufacturers are doing their best to make in mass production what is really a laboratory device."[6] The inherent fragility of the tubes made them difficult to rely on. Most vexing was the peculiar inaccessibility of the tube's electronics, which, thanks to the vacuum surrounding them, could not be poked, prodded, or tested while under operation. It was to perform such basic research on tube electronics that Litton recruited noted tube engineer Siegfried Hansen from Hughes.

Siegfried "Siggy" Hansen had had a long career, including a role at General Electric in the invention of the modern television picture tube. His approach at Litton was unprecedented, but simple. Traditional tube research involved the time-consuming construction of multiple tubes, each laboriously hand-built and sealed before testing. As an alternative, Hansen proposed to make a vacuum chamber large enough for the bulbs' electronics to be manipulated and tested by a technician while under electrical load.[7] With a very different problem than an airborne pressure suit, and without the attendant weight and mobility requirements, Hansen produced a very different kind of vacuum outerwear.

THE LITTON MARK I "VACUUM" SUIT

For the legs, torso, and helmet of the vacuum technician's suit, Hansen borrowed from different existing technologies. The feet, legs, and groin of the suit were borrowed from a high-altitude flight suit—although when exposed to the near-vacuum of the test chamber, they ballooned and became immobile, restricting the operator to a shuffling hop. The torso and helmet of the suit were made like a deep-water diving suit, manufactured in aluminum with a Lucite window on the helmet. To allow the operator to don the suit, an opening was placed diagonally around this chest tube, resembling a bandoleer of bullets with its row of massive bolts. Heavy restrictions on the operator's large-scale movement were deemed acceptable for the suit's laboratory role. Hansen's biggest problem, however, came in the suit's gloves and arms, where the ability to perform delicate manipulations and measurements was essential.

While the spacesuits developed from Hansen's vacuum suit would be termed "hard" (in comparison to the all-fabric Mercury, Gemini, and Apollo suits), this was, in terms of the actual innovations involved, somewhat of a misnomer. Instead, the great advance in mobility came through developing a structure for the moving arms of the suit that ingeniously fused the rubberized fabric of the high-altitude pressure suit with rings of rigid metal—recalling the constructivist costumes designed by the Bauhaus's Oskar Schlemmer for his *Triadic Ballet*. This was perhaps not a total coincidence; joining the team in 1958 was an Air Force engineer (and later Litton manager) William Elkins. He had studied industrial design in postwar Chicago on the GI Bill, and came into regular contact with the city's Bauhaus émigrés.[8]

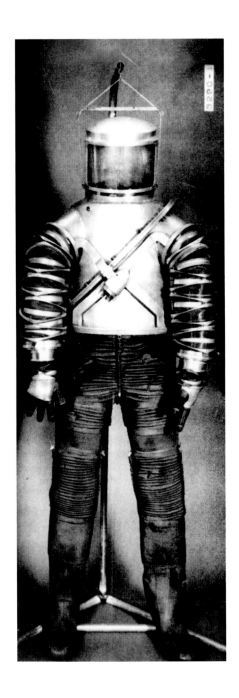

15.1

The Litton Mark I suit without protective covering over its "cardonic" convolute arms.

Ironically the highly mobile arms of Litton's vacuum suit had less in common with the tomato worm bellows developed for flight suits in the 1940s (Hansen and his team had tested these and found them inadequate for the full vacuum of their chamber) than with the pleats and gathers of more traditional clothing. The pleats engineered by Hansen, however, were a composite of hard, metal rings, sliding across one another within tight folds of fabric. As the engineers discovered, these composite pleats moved very easily over each other as the suit changed geometry—but there still remained the problem of the enormous expansive strain put on the mechanism when the suit was filled with a pressurized atmosphere. To keep the pleated, or "rolling convolute" joints from expanding as the suit filled with air, they added an external skeleton of metal rings, connected to each other and bolted, through metal and fabric, to each joint, or pleat. The compression of these hinged rings allowed the assembly to resist the expansion that would have otherwise been caused by increasing internal pressure.

An even greater hybrid of elements was used when Hansen's team reached the most critical part of the manipulator's suit, the hands. An aluminum palm was matched with flexible rubber fingers, which, while they seemed to resemble the segmented rubber of early flight suits, were restrained from ballooning by internal nylon tapes to which the rubber was chemically bonded (an arrangement similar to the rubber convolute used by ILC). While limited in their feel and dexterity, the composite gloves provided a greater ability to work in a vacuum than any previous flight suit glove. The Mark I Litton vacuum suit, as it was to become known, was ready for work in the vacuum chamber as of mid-1957. The rapid obsolescence of vacuum tubes, however, threatened to make the suit worthless as well.

Even as Tex Thornton was purchasing Charles Litton's vacuum tube business in 1953, *Fortune* magazine proclaimed the "year of the transistor." The new device, invented at Bell Laboratories in the late 1940s, relied on quantum effects at the interface between two materials to produce the same electronic effects as vacuum tube diodes and triodes. A small Houston company, Texas Instruments, stunned the emergent industry by displaying the first, highly stable, silicon transistors in 1955.[9] The late 1950s—the very period when Hansen and his technicians were perfecting the space of their vacuum chamber and technician's suit—was precisely when the technical details of transistor production were successfully tackled, not only at Texas Instruments but also in the small cluster of companies that resulted from the departure of William Shockley (supervisor of the original transistor team at Bell Labs) for a Quonset hut near his parents' home in Palo Alto. (In 1969, the area's Chamber of Commerce would coin the term "Silicon Valley.")

SPUTNIK

The final impetus for U.S. transistor development came from the launch of Sputnik in October 1957.[10] Ironically, the Soviet satellite was a miniature analog to Litton's rigid pressure suit. Unlike U.S. satellites, which used transistors from the outset, Sputnik had a pressurized atmosphere, the main purpose of which was to prevent damage to the vacuum tubes that constituted its electronic equipment. As the vacuum of space was inevitably more empty than the vacuum achieved by Russian tube-making machinery, the Soviet "vacuum" tubes would otherwise have exploded.[11] The alternating radio signal of the satellite became a signature of the cultural impact of Sputnik—

the "chilling beeps" editorialized by *Time* in its coverage of the launch ("Red Moon over the U.S.").[12] More than a simple beacon, the duration and timing of the beeps contained coded information about the status of the satellite's pressurized internal environment.[13]

The potential of Litton's hard vacuum suit for space exploration had occurred to the Air Force as early as 1954 (when what became the Mercury program was already being considered). For fear of congressional ridicule, however, the contract supporting suit development issued to Litton by the Air Force in 1954 forbade any mention of the term "space" or "suit," instead specifying the innocuous "manipulator station." Sputnik changed all of this, and had the Air Force touting its research to all comers as evidence of its (supposed) ongoing and extensive preparation for the space race. The newly dubbed Litton Mark I spacesuit appeared throughout the popular press as one of the only tangible images of U.S. space research.[14] Litton—already expert at media rela-tions—dubbed Hansen's facility the "Space Laboratory," and cemented its image on the leading edge of corporate technological enterprise.

15.2
The Litton Mark I suit on the cover of *Look* magazine,
December 10, 1957.

RX FOR SUIT DESIGN

Further support for Litton's suit research arrived with the enormous budgets of Apollo. The suits worn on the surface of the moon, NASA recognized, would need to be very different from the backup bladders worn by Mercury astronauts. The durability, mobility, and calculable engineering merits of Litton's hard suits proved immediately appealing. From 1963 to 1966, Litton's suit program—now under the direction of Bauhaus-influenced Bill Elkins—produced multiple iterations of "RX" experimental spacesuits. Unlike the Mark I suit produced under Air Force contract, these were to be the first spacesuits produced by the firm that would be so termed from the outset. They were designed precisely to solve the problems of durability, mobility, and protection that were anticipated in Apollo's moon landing goal. The RX in their names—which officials were told stood for "rigid experimental"—was actually the result of an assertion by NASA suit research director Joe Kosmo that the suits were a "prescription for the ills" of Apollo pressure suit design.[15]

Beautiful and rigidly armored, the first RX-1 and RX-2 Litton suits provoked a flurry of press releases from NASA. "Nothing new under Ye Olde Sun?" one image was captioned, contrasting the innovative armor of the suits to the historic heroism of medieval costume.[16]

The substantial innovation of the RX suits was a new, flexible joint. In Litton's Beverly Hills Laboratory, Elkins had conceived of a way to make the metal-and-fabric pleats of the Mark I suit slide over each other instead of sitting end-to-end. The resulting "stacked convolute" increased the strength and number of layers over sensitive parts of the suits' structure while allowing most of the legs, arms, and torso to be protected by rigid tubes instead of (relatively) fragile joints.

The RX-1 suit deployed these new techniques in an aluminum carapace over its wearer's head, torso, and lower legs, leaving the difficult geometry of the hips and groin to be dealt with by a soft rubber bladder. Seeking to improve on the suit's innovations, Litton disassembled the RX-1 and reassembled its successor RX-2 from the component parts—this time crafting a completely rigid enclosure. As of the RX-2, the Litton suits became fully "hard" in another, more subtle, respect— their interior volume did not change as the suit was moved by its wearer, making movement in the suit much easier—but putting enormous demands on the suit's engineers to fully resolve human movement into a series of mechanical joints and motions.

Like subsequent Litton prototypes, the RX-2 was not clothing, but emphatically a rigid enclosure. Instead of draping, circling, and supporting the body with a tight air bladder—against which the astronaut pressed his body to move the suit—the Litton suits did not touch the body of the wearer at all, except for the soles of his feet and the inside of his (flight-suit-derived) gloves. Engineers especially were said to prefer the sensation of a suit that set around, not on, the body.[17]

Like a vehicle, but unlike clothing, the suit even had a seat—a repurposed bicycle saddle on which the astronaut's body was supported. (The vertical adjustment for the seat appeared as a dial on the suit's groin.) The bulk of the suit's weight hung from a harness over the shoulders.

While the aluminum RX-2 of 1964 appears Frankenstein-like, the RX2A produced the following year presents a streamlined aspect; its compact shape and clean aesthetics alone were cited as a reason to support the suit's use by devotees inside and outside the military-industrial establishment.[18]

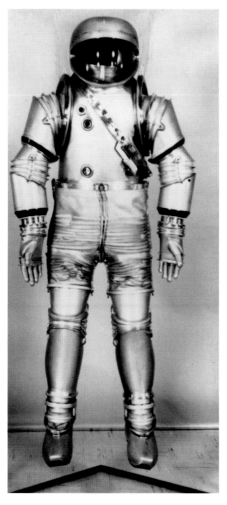 

15.3
NASA press release, "Nothing New under Ye
Olde Sun?," June 26, 1964.

15.4
The RX-2 in a NASA photograph from 1965.

15.5
The RX2A, 1965.

Across multiple generations of Litton suits, which would include development contracts for an RX-3, 4, and 5, these geometries would be increasingly refined. Initially, seeking an effective shape for the RX-1 suit's rigid shoes, the Litton engineers purchased a $4.50 pair of clogs and (after sanding off the pointy toes) cast them in hollow aluminum. Later versions of the suit involved precision molds made from complex shapes, themselves derived from elaborate mobility testing inside earlier prototypes. The suit surface itself shifted from solid metal into a lightweight composite of aluminum, fiberglass, and hexagonally arranged insulation.

A 1967 study by NASA's Human Engineering Section found the the RX-3 suit equal to or better than the ILC suits being developed in parallel in several important tests.[19] As a result, NASA saw the RX suits as a crucial backup to the soft suits should they fail in either testing or—it was feared—actual use.[20] While their forms were more difficult to customize to astronauts' individual body sizes than a soft suit's, the RX-3 and 4 suits introduced a system of sizing that could use a standard series of shells to assemble suits for 90 percent of astronauts. And while it was feared that the suit's mechanical joints might become fouled by lunar dust and grit, in one other aspect they were more than equal to the latex bladder suits made by ILC: unlike the balloons at the core of the soft suits, the Litton suits did not decrease in flexibility as pressure inside of them increased.

NAME OF COMPONENT	SIZE
1. HELMET ASSEMBLY	STANDARD
2. UPPER TORSO ELEMENT	I, II, III, IV
3. SHOULDER JOINT	STANDARD
4. UPPER ARM ELEMENT	STANDARD
5. UPPER ELBOW ARM ELEMENT	I, II, III, IV, V
6. ELBOW JOINT	STANDARD
7. FOREARM ELEMENT	I, II, III, IV
8. WRIST JOINT AND ROTARY SEAL	STANDARD
9. GLOVE SHELL	I, II, III, IV
10. LOWER TORSO ELEMENT	STANDARD
11. WAIST RING	STANDARD
12. PANTS	I, II, III, IV, V, VI
13. HIP JOINT	STANDARD
14. THIGH ROTARY SEAL	STANDARD
15. THIGH ELEMENT	I, II, III, IV, V
16. KNEE JOINT	STANDARD
17. CALF ELEMENT	I, II, III, IV, V
18. ANKLE JOINT	STANDARD
19. BOOT	I, II, III
20. EV VISOR	STANDARD

15.6

The modular sizing system developed for the RX-3 and 4 series of suits.

15.7

The ejection system proposed to mate the RX-3 hard suit with the Manned Orbital Laboratory's Gemini Capsule, including a rocket-powered ejection mechanism.

SOLDIERS IN SPACE

While it was acknowledged that a failure of the RX suit's single layer would be catastrophic, the image, as well as the increasing substance of its greater durability, provoked interest outside of NASA as well. Taking a particular interest were the military planners of America's first proposed military outpost in space, the Manned Orbiting Laboratory (MOL; pronounced "mole").

The MOL was an Air Force project, for which 14 astronaut-soldiers were recruited to man an orbital reconnaissance and defense station. In orbit for 45 days at a time, the station was to be built around a repurposed Gemini capsule and Titan ICBM, which would lift an extended-use crew compartment as well as the original Gemini capsule. As a military outpost instead of a civilian space facility, the MOL was subject to a range of design requirements quite different from the country's more public space effort—including consideration of the astronaut's need to eject from the Gemini capsule if under attack. For this eventuality, Litton proposed a specially modified RX-3 suit, whose shoulders and hips locked, through a connector, to a rocket backpack that would allow separate descent to earth.

A system of rapidly inflating bladders, similar to the airbags of a modern car, was designed to inflate inside the suit to immobilize the ejecting astronaut and protect him from shock.

LOR AND LAYERS

Despite its likely success in a space environment, the Litton suit turned out to be ill adapted to important realities of the space race. The first of these involved a debate that raged inside NASA until June of 1962 about the precise strategy to be used to reach the moon.

The first landing strategy considered for Apollo—advocated by missile manufacturers like Martin Marietta, Douglas, and General Dynamics—was seemingly the most straightforward: using the most enormous rocket then imagined, the Nova, NASA would launch a single spacecraft from earth with enough force to land a vehicle directly on the moon, complete with enough fuel to blast off from the moon and return directly to earth. Termed "direct ascent," the method would have resembled that of classic moon journey narratives from Verne to Tintin. The Nova rocket needed would dwarf even the Saturn V, and, not incidentally, be produced by the same companies supporting the strategy.

The second proposal—advocated by Wernher von Braun and his group at the Marshall Spaceflight Center—was termed "Earth orbit rendezvous" (EOR). In this scenario, equipment and fuel for a lunar mission were to be ferried to earth orbit by several Saturn V rockets. The lunar vehicle would be assembled above the earth, then leave orbit to land directly on the moon. While the proposal involved complicated orbital maneuvers to assemble craft, crew, and fuel, these would take place only hundreds of miles above the earth's surface—allowing, it was imagined, for much double-checking before man and machine were blasted a quarter-million miles further away.

Until the very last moment, it was assumed by all involved in the process—including NASA's own leaders—that one of these two methods would be chosen. At the last minute, however, a third option emerged. "Lunar orbit rendezvous" (LOR) was a complex strategy to get to the moon that boosted far less weight into orbit than either of the first proposals. Using a range of technologies already under development in 1961, the proposal placed its most exacting demands not on the massive rocket booster itself, but on the capsules and astronauts at its very tip.

LOR's delicate combination of multiple vehicles, docking and undocking in lunar orbit, seems in retrospect inevitable. Yet at the time it seemed as strange as the purpose-built vehicle it spawned—the spidery lunar lander. Without backing from either von Braun—whose proposed EOR would guarantee a massive production of the Saturn V boosters he and his team were developing—or the missile manufacturers—to whom the idea of a massive, expensive new rocket necessary for direct ascent was mouth-watering—the proposal languished. In the end, the ticking clock itself decided; LOR, it became clear, was likely the only strategy that could accomplish a lunar landing before 1970. While the most complex in its orbital logistics, it was in its material substance, and consumption of resources, the simplest. It would use only one Saturn V vehicle for launch, would not demand the development of entirely new technologies, and had the further benefit of separating two of the most difficult tasks of the mission—the launch and landing compartment for the crew, and the device to land on and take off from the moon—into two separate physical devices, whose design and organization could proceed separately. One significant result of the LOR decision, however, was an extreme limitation on weight and volume in the lunar vehicles.

Launched by a single rocket, the two vehicles of Apollo—Command Module and Lunar Excursion Module, or LEM—became marvels of miniaturization. The inside of the Command Module (in which some crews spent almost two weeks) was the same size as a 1969 Volkswagen van. The LEM was much smaller. The size and weight restrictions forced by the LOR decision meant that every component of the mission was pared down and made to do double duty. The astronauts would take only one pressure suit each into orbit; this would serve both to protect them against depressurization of the Command Module on launch and reentry, and to protect two of their number on the moon's surface. Furthermore, these pressure suits would have to be stowed for inbound and outbound journeys inside the Command Module, where every cubic inch or ounce saved meant a greater number of lunar samples returned.

The logistics of rocket flight itself further intervened to trim weight from the Command Module; every ounce trimmed at the tip of the rocket saved thousands of pounds in fuel in the massive launch stages. Despite a proposal by Litton engineers to stow the suit's parts inside each other like a Russian doll, the softness, squishiness, and stowability of the flight-suit-derived alternatives were better adapted to the micro-efficiencies of the LOR strategy. The fact that ILC's suits were defying expectations of soft suits in performance, comfort, and durability was the deciding factor; Litton was not considered a serious competitor for the first, "Block II" series of lunar flights.[21]

15.8
Litton's RX-5 being tested on a simulated lunar surface in 1968.

While the LOR strategy gave NASA a viable way to reach the moon by 1970, it did not give Litton the prime space-borne market for its exhaustively marketed suits. Then, one by one, alternative opportunities disappeared; while military strategists approved the RX suits' durability and strength for the planned MOL station, weight and stowage concerns similar to those disqualifiying the suits from Apollo meant that Hamilton Standard pressure suits were chosen for MOL's initial occupants. Soon after, in 1969, the MOL program itself was canceled, its orbital darkroom rendered obsolete by the electronic sensors of newer spy satellites.[22]

The hopes that Litton's RX suits would fly on later Apollo missions were dealt their first blow in 1966, with the acceptance of ILC's "Omega," or A7LB, suit for missions 14–16 (which involved longer stays on the moon and the use of the Lunar Rover). A subsequent hope of Litton engineers (as well as their supporters in NASA's Crew Systems Division) was to have their suits used on even more extended Apollo landings. Missions 17–20 were planned to involve more arduous exploration, longer stays on the surface, and a new round of contracts for equipment such as pressure suits. These hopes died when missions 18–20 were canceled in 1970. (The Litton suits competed in each case with offerings from AirResearch, a division of the Garret Corporation, now Honeywell. William Elkins had moved from Litton in 1966, and led AirResearch efforts from 1967.)

Litton suit technologies finally played a role in Apollo when constant-volume arm assemblies were used by workers to manipulate lunar samples after Apollo 11; soon after, however, these assemblies were replaced with a David Clark-produced bladder-and-link-net design. Litton's RX suits, however, were to have a central role in one remaining lunar panorama. There, as at their inception, they would form the public image of spaceflight's future.

THE "BUNION PAD"

Buckminster Fuller's 1967 U.S. Pavilion at the Montreal world's fair had wowed the world with its "synergistic" geodesic geometry. U.S. space exhibits, including the fully deployed parachutes of the reentry capsule of the Command Module, chiefly occupied its large interior. The subsequent U.S. Pavilion at Osaka's 1970 fair has received less historical attention.

Partly, this is due to the differences in scale between the two structures. While the original 1970 proposal (by Chermayeff & Geismar, with architects Davis, Brody) included a giant, inflatable moon surrounding an experimental mediathèque, budget restrictions on the Vietnam-era government forced substantial cuts in size and scale.[23] "There will be many large, lavish structures at Expo 70 in Osaka," reported *Architectural Record*, "but the U.S. Pavilion will not be one of them."[24] The resulting, low-profile design, derided by *Architectural Forum* editor Peter Blake as the "biggest bunion pad ever,"[25] was the largest air-supported structure then built. Deliberately constructed of the same Beta Fiberglass fabric as the AL7 spacesuit[26] (although without the expensive Teflon coating on each thread), the pavilion contained exhibits on photography, painting, architecture, folk art, and sports equipment.[27]

Commanding the greatest area in the pavilion, however, was the exhibit of U.S. space technology, which included capsules from Mercury 7, Gemini 12, and Apollo 8; a moon rock brought back by Apollo 11 was installed on a tall pedestal. The interior of the large space exhibit was densely populated with "spacewalking" spacesuits, on a timeline from Mercury to Apollo.

15.9

Davis, Brody with Chermayeff & Geismar,
the U.S. Pavilion, foreground, at the Osaka
world's fair, 1970.

15.10 (following pages)

Davis, Brody with Chermayeff & Geismar,
U.S. Pavilion, Osaka world's fair, interior
showing space exhibit, 1970.

15.11
Davis, Brody with Chermayeff & Geismar,
U.S. Pavilion, Osaka world's fair, interior showing
space exhibit and Litton RX-1 and RX-5, 1970.

The final tableau, however, was taken to represent the future of space exploration; a massive lunar panorama containing the Litton RX-2 and RX-2A prototypes.

By the time they appeared in a simulated panorama in Osaka in 1970, the Litton hard suits were an organizational and institutional footnote. Yet, in the same way that Litton's corporate image was later criticized for a victory of style over substance (see Layer 20), the shiny, silver surface of the RX-2, and even the glossy white, segmented RX-2a, were well equipped to project a clear, solid image of America's future in space; certainly more so than the softer reality of the AL7 suits deployed elsewhere in the exhibit.

Like the Bauhaus costumes to which Litton's William Elkins had been instrumentally exposed, however, the Osaka panorama was a utopian vision. With their smooth curves and exoskeletal frameworks, the suits had the potential to delight and dazzle—but not to survive the soft realities of the space race.

Notes

1. William B. Harris, "Litton Shoots for the Moon" *Fortune,* April 1958, 114–119, 206, 208, 210, 212.

2. John A. Byrne, *The Whiz Kids: Ten Founding Fathers of American Business—and the Legacy They Left Us* (New York: Doubleday, 1993), 33.

3. Ibid., 34.

4. Ibid., 88–108.

5. U.S. Patent 879,532, February 18, 1908 (application filed January 29, 1907).

6. Dr. Ralph E. Myers (Vice-President, National Union Radio Corporation), "How to Get the Most out of Vacuum Tubes," *New York Times*, September 21, 1930, R10.

7. Gary A. Harris, *The Origins and Technology of the Advanced Extravehicular Space Suit*, American Astronautical Society History Series, 24 (San Diego: AAS Publications, 2001), 120.

8. Nicholas de Monchaux, interview with William Elkins, June 2, 2009, audio recording.

9. See Michael Riordan and Lillian Hoddeson, *Crystal Fire: The Birth of the Information Age* (New York: W. W. Norton, 1997), 209–219.

10. Ibid.

11. See M. K. Tikhonravov, "The Creation of the First Artificial Earth Satellite: Some Historical Details," first presented in Russian at the 24th International Astronautical Congress at Baku, Azerbaijan (then part of the Soviet Union); translated in 1994 by Peter A. Ryan and published in the *Journal of the British Interplanetary Society* 47 (1994): 191–194.

12. *Time*, October 14, 1957.

13. Paul Dickson, *Sputnik: The Shock of the Century* (New York: Walker, 2001), 129.

14. Lillian D. Kozloski, *U.S. Space Gear: Outfitting the Astronaut* (Washington, D.C.: Smithsonian Institution Press, 1994), 148; William Smith, "Space Pioneers: Medics Push Lab Tests of Men, Material for Future Rocket Flights," *Wall Street Journal*, December 4, 1957, 1.

15. Joe Kosmo, interview by Nicholas de Monchaux, Houston, May 1, 2006, audio recording.

16. Caption to NASA Image 64-Apollo-104, Record Group 255-G, National Archives Still Picture and Photograph Collection, National Archives at College Park, College Park, MD.

17. Elkins interview, June 2, 2009.

18. Kosmo interview, May 1, 2006.

19. NASA Document HES-67-13, May 3, 1967, "Evaluation of RX-3 Hard Suit," prepared by Joe E. Reed and Don E. Kirkpatrick, Johnson Space Center Archives, University of Houston-Clear Lake, Center-Spacesuits Collection, Box 7.

20. Kosmo interview, May 1, 2006.

21. "Block I" Apollo flights had begun and ended with the Apollo 1 launch fire in January of 1967. Modified Gemini spacesuits, termed A1C by David Clark Co., were decided on in 1965 for the early Apollo flights to allow additional time for (then contentious) Lunar EVA suit development. By the time manned Apollo flights resumed in late 1968, ILC suits were ready for use.

22. Tex Thornton himself was mentioned as a possible successor to McNamara at Defense ("Johnson Vows No Loss of War 'Momentum' as He Announces Departure of McNamara," *Wall Street Journal*, November 30, 1967, 5), but it was Clark Clifford who got the job and canceled MOL as one of his first, budget-saving measures.

23. An even more moon-themed entry came from Philip Johnson, who proposed setting fairgoers loose on a giant, simulated lunar crater. "The U.S. at Osaka," *Architectural Record*, October 1968.

24. Ibid.

25. Mark Dessauce, *The Inflatable Moment: Pneumatics and Protest in '68* (New York: Princeton Architectural Press, 1998), 145.

26. David Geiger, "U.S. Pavilion at EXPO 70 Features Air-Supported Cable Roof," *Civil Engineering—ASCE*, March 1970.

27. David Geiger, "Low Profile Air Structures in the USA," *Building Research and Practice*, March/April 1975, 80–87.

Layer **16**   "WE'VE GOT A SIGNAL!"

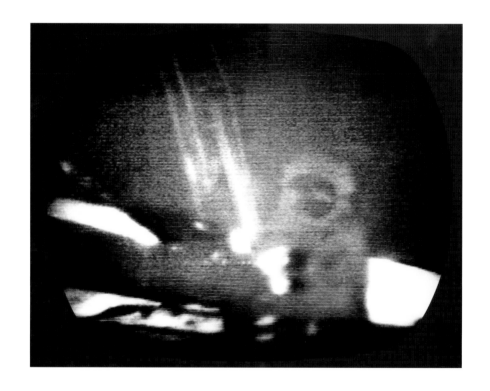

16.1
Neil Armstrong stepping onto the lunar surface.
Screen capture from television broadcast.

With an announcement by Mission Control—"We've got a signal!"—the first lunar television picture was broadcast. Above the astonishing caption, "Live from the surface of the moon," two black-and-white shapes resolved themselves into a blurry, upside-down view of Neil Armstrong, paused on the Lunar Module's ladder. He had paused, in fact, to maneuver a complex apparatus that pulled a television camera out from the Lunar Descent Module, and train it on himself. Armstrong waited for almost a minute on the ladder while NASA technicians adjusted the image, and only then proceeded to step onto the lunar surface. In the CBS broadcast of the landing there followed a reaction shot of Walter Cronkite, tears streaming down his face.

Edwin "Buzz" Aldrin, descending after Armstrong, brought with him a mobile version of the same camera on a long umbilical line. After some problems with tangles in the cable, this second camera was brought to the lunar surface. Neil Armstrong took it with him on one of his furthest walks from the landing craft, the distance of a short New York block. The purpose of the walk was to set the camera on a tripod at a vantage point that would show the rest of the brief program of extravehicular activity. Armstrong, who could not see the TV image, was instructed by ground control where to place the camera and at what angle to frame the lunar panorama.

The worldwide audience for the Apollo broadcast was estimated at 528 million—nearly 15 percent of the world's population. While only a small cadre of party elite watched the broadcast in the Soviet Union, 95 percent of televisions in the United States were tuned to the broadcast— a large majority to the "superior coverage" of CBS.[1]

The CBS broadcast was at once a throwback and a flash-forward. On the one hand, it recapitulated a relationship—complicity, even—between media and government that was, by the end of the 1960s, already under siege. On the other hand, even as a relic of a now-extinct media monoculture, the broadcast contains the germ of much of contemporary television culture.

THE MEDIA EXECUTIVE

Despite the talk of "all mankind," Apollo was a U.S. government enterprise, and of the presidency in particular. As such, it is quintessentially representative of the collaboration between presidents and television that characterized much of postwar reality.[2] Yet this complicity would undergo a remarkable transformation from 1960, when, as detailed in Layer 10, John F. Kennedy's command of media led directly to the decision to fund Apollo.

If Kennedy and Nixon represented the alpha and omega of media fluency—the dark, distracted Nixon punch-drunk under the TV lights, Kennedy cool and tan—then Lyndon B. Johnson was an uneasy intermediary, presiding over the transformation of U.S. media from willing boosters of executive power to, at least occasional, critics. No event was more important in this transformation than Johnson's pursuit of the war in Vietnam. Its defining juncture, the Gulf of Tonkin attack, was arguably a staged television event, and Johnson was relentless in appropriating TV time for the war's explanation—and defense.[3] At the same time, it was the media's reporting on the realities of Vietnam—first in print, and later in broadcasts by figures like CBS's Harry Reasoner—that turned public opinion irrevocably away from Johnson.

Looking back, this shift in the mediated presidency seems inevitable. It was not. Even at the beginning, it was Johnson's longtime friendship with Frank Stanton, president of CBS,[4] that had helped lay the groundwork for the 1969 lunar broadcast; friendship with Stanton would bring years of positive coverage to the president. As previously noted, Stanton was a member of the small group that advised Johnson on his memo outlining responses to Gagarin's orbital flight in 1961, and who helped emphasize the television power of the image his network would come to broadcast. This close relationship between Johnson and the media continued after Kennedy's death, into his own administration. "You always come through, Frank," Johnson wrote Stanton in 1964, "and I am grateful for it." The influence of Johnson's power on America's most trusted news source was systematic. CBS's impact was even more particular; Stanton personally designed and rebuilt Johnson's Oval Office desk to help the president be more telegenic.[5]

Within Stanton's CBS, the tension between entertainment and news, so characteristic of our own era, was first appearing. In 1966 William Paley, Stanton's sole superior at the network, went before a stockholders' meeting and announced that, had it not been for unscheduled news events such as Gemini flights and the death of Winston Churchill, the shareholders would have received six cents more in dividends. The same year—and five years after Edward R. Murrow himself had left the network—longtime producer Fred Friendly resigned in protest at the commercialization of news coverage. He subsequently declined an offer to come to the White House as Johnson's television advisor.[6] The tension between management and reporters over news coverage was fueled by increasingly critical coverage of Vietnam, a crescendo that culminated in 1968 with Cronkite's televised denunciation of American involvement.

Against this background, the 31-hour CBS broadcast "Man on the Moon: The Epic Journey of Apollo" was a marvelous throwback to the days of Kennedy and Camelot—not just the effectiveness of the lunar television image that Kennedy had visualized in 1961, but the collective engagement of media, government, and citizenry in a single, glorious, TV image.

SYSTEMS AND S-BANDS

The 1969 lunar broadcast was a moment of complicity, and even synthesis, between the organizational fabric of NASA and that of news organizations like CBS. In organizational terms, however, this collaboration was hard-won. Despite Apollo's origins as a media event, the organizational logic of NASA came to threaten the very essence of Kennedy's goal—the transmission of live television from the moon.

NASA's systems engineers, insulated from the media origins of Apollo, and contractually responsible for the safety of its astronauts, had in 1961 allocated a mere 700 Mhz of bandwidth from the lunar surface for television broadcast. This was a fraction of that apportioned, for example, to telemetry, or monitoring the astronaut's heartbeat or temperature. Once this 1961 decision had been imbedded in the systems architecture, the so-called S-band transmission guidelines might as well have been carved in stone.

It was only in 1964, after several years of grinding away at the means and mechanics of Apollo, that the end—the television image—was again considered. Realizing the ramifications of the S-band apportionment, NASA officials were furious at the thought that the 1961 allocation might jeopardize a broadcast from the moon's surface. A "desperate" Jim Webb asked both RCA and Westinghouse to propose a mechanism whereby a television picture, and perhaps even the increasingly ubiquitous color signal, might be squeezed into this small space: one-tenth the size of an earthbound black-and-white signal. The solution, devised by Westinghouse engineers, was a radical adaptation. Instead of the 640 lines of a conventional American television, the lunar camera would produce a picture with one-fourth the resolution. And while the NTSC broadcast reaching American living rooms in 1969 refreshed itself every one-sixtieth of a second, the lunar broadcast would be a choppy three times slower.

Squeezing itself around the edges of the more "mission-critical" technical transmissions, the lunar television signal would be subject to a final indignity—adapted to the special circumstances of its tiny bandwidth, it was in no way compatible with earthbound cables or transmission systems. After experimenting with more complex procedures to translate the Westinghouse camera output to earthbound broadcasts, the engineers settled on a simple, if low-tech, solution: the Westinghouse signal would be shown on a monitor at the various ground stations charged with receiving the S-band broadcasts, and be sent further from a standard TV camera pointed at the screen.

As a result of its elaborate transmission, the final lunar image was a study in chiaroscuro, shoals of slowly moving light and dark. Due to the difficulty of adjusting exposure against the contrast-laden environment of space, the image had to be continuously adjusted by hand.

HAL 10000

If the actual transmissions from the surface of the moon were hamstrung by conflicts between the systems of NASA and the infrastructure of television, the overlap between NASA and broadcasters would reach its apotheosis in one of the most futuristic components of CBS's Apollo coverage— its complex simulations and multilayered visual effects. In the absence of a live feed preceding Armstrong's exit to the lunar surface, CBS viewers were treated to hours of simulations and staged reenactments of events in lunar broadcast. As the Lunar Module *Eagle* descended slowly to the lunar surface, gliding on the rails of its parabolic trajectory, for example, the viewers of CBS enjoyed a bird's-eye view from above the spidery lander, and the cratered landscape rolling below.

Literally rolling, as it turns out. The landscape was an enormous, latex-and-fabric belt, rolling away from a special-effects camera. Electronics stitched the moving image together with images of an LEM model, a rising earth, and electronic text to indicate minutes to landing. Immediately prior to Armstrong's live feed, the CBS audience watched actors simulating moonwalk preparations on a Long Island soundstage. The scene was complete with a Lunar Module rented from Grumman Aerospace for the purpose, and a moonscape made of mining slag.

Planning for the 31-hour CBS broadcast had gone back half a decade. At the same time as NASA had started to realize the limitations on its own lunar television technology, CBS itself had seen that it could not rely for its production on materials provided by the government. Wary of depending on the single, potentially unreliable S-band signal, and dismissive of NASA's own attempts at supplementary visual materials ("second-grade stuff," one engineer scoffed), CBS embarked on a massive investment in its own broadcast technology, guaranteeing (the network hoped) a successful broadcast. There were two direct results, each tangible in today's media landscape.

16.2
A still from the CBS lunar broadcast. © CBS.

The first derived from a growing complicity not just between NASA's press office and CBS (a specially rented telephone line ran from the desk of Paul Haney, NASA's press spokesman, to CBS's broadcast studio),[7] but between CBS and NASA's efforts in mission simulation. While such relationships already characterized CBS's earlier space broadcasts, for Apollo they went into overdrive. CBS spent hundreds of thousands of dollars on visual effects for the lunar broadcast, mostly from the same sources and contractors engaged by NASA itself. These included Grumman, North American Aviation, and Aero Service Corporation, a division of Litton Industries that produced transparencies for use in NASA simulators.[8]

At the same time, CBS borrowed technology in noise reduction and signal processing to bring the superimposition of these separate visual elements (using colored-screen, or chromakey, technology) to new levels of visual sophistication.[9] Especially compared to other programs of the time, the lunar broadcasts were a visual feast, progressing from shots of Walter Cronkite in front of the Saturn booster to multilayered, special-effects-ridden shots inside the studio. Many of these made extensive use of the simulators and models commissioned by CBS.

But most of the intriguing complexity of the broadcast was hidden from viewers. For the shot of the *Eagle* lander in its approach trajectory, at least four images were superimposed—a miniature model earth, a model lunar lander, a tracking shot of the latex-conveyor lunar surface, as well as a subtitle, "CBS News Simulation." In other shots of the broadcast, up to four images appear separately on the screen at once, from Houston, Florida, and New York studios.

Exemplary of CBS's vast coverage infrastructure was the massive set to which Walter Cronkite decamped after broadcasting from Apollo 11's launch at Cape Kennedy. Incorporating an enormous, raised studio desk (below which sat a researcher, ready to hand Cronkite scripts to accompany each broadcast segment), the massive apparatus itself integrated a dozen screens and borrowed technology developed for Houston's Mission Control to show the progression of orbital trajectories and mission timing. With its fiberglass shells and flashing colored screens, the desk was called "HAL 10000" by CBS staffers; this was not just for its literal resemblance to the smooth, space-age aesthetics of Stanley Kubrick's 1968 film, but for the fact that it had also been designed by Douglas Turnbull, visual effects supervisor for *2001*.

16.3

The latex rubber moonscape belt used in CBS simulations. The belt was cast from the same mold as the landscapes of NASA's Apollo simulators. Note model of earth at right.

16.4
Walter Cronkite seated at HAL 10000.
Cronkite spent 17 hours continuously
broadcasting on July 20, 1969. © CBS.

CLIP BANKS AND CABLE

If one kind of future was anticipated by the CBS broadcast's simulated content, another was forecast by its form.

On September 2, 1963, CBS became the first broadcast news organization to schedule half-hour evening news broadcasts, expanded from the previous slot of 15 minutes. When, around the same time, CBS News's Special Events division first examined the prospect of covering weeklong Apollo missions, and multiple days of continuous coverage from the lunar surface, the organization had "no idea" how to fill such an unscripted expanse of time. With due deliberation, a plan was developed. Starting as early as 1964, CBS assembled footage for the Apollo landings—not of the actual events, of course, but related material, including simulations, background stories, and related interviews that might help fill the time. Instead of being arranged linearly, as with a scripted broadcast, these clips would, for the first time, be placed in a nonlinear "bank," accessible on demand as the coverage was broadcast.

While the "clip bank" was being prepared, CBS lined up hundreds of potential commentators and critics of the anticipated events—not just stalwarts such as Wally Schirra, or *2001* author Arthur C. Clarke, but figures like Gloria Steinem and Ira Magaziner (leader of the recent student revolt at Brown), who could be expected to add a seasoning of controversy to the splendid moment. (As it transpired, Magaziner and Steinem only partially filled their scripted roles: "I am very excited, certainly," Magaziner allowed.)[10]

The resulting form of CBS's 24-hour Apollo coverage—talking heads and roundtable discussions, manufactured dialogs interspersed with prerecorded segments and occasional break-ins for breaking news—was as unfamiliar to 1960s audiences as it is ubiquitous today. The producer of CBS's broadcasts, Robert Wussler, would take the lessons learned during the Apollo coverage to found the Cable News Network (CNN) with Ted Turner in 1980; this became the first of more than seventy news outlet networks currently providing 24/7 coverage using the format designed by Wussler for Apollo.

CAPRICORN ONE

A final legacy of the mediated, simulated lunar broadcast became the most infamous theory surrounding the moon landing—the idea that all of Apollo was a hoax. One of the first to imagine the conceit was CBS reporter Peter Hyams. "There was one event of really enormous importance," Hyams later reflected, "that had almost no witnesses. And the only verification we have ... came from a TV camera."[11]

Yet if the lunar conspiracy theory can be dated to the media circus of Apollo, its success should be ascribed to a much more somber spectacle—the TV drama of Watergate. While the *Washington Post* famously broke Watergate in old-fashioned newsprint, it was the two-part special feature by CBS in 1972 that is credited with keeping the story alive through the winter of 1972 and 1973.[12] Not originally considered a "TV story" by some at CBS, it was to become such with the riveting Senate and House hearings throughout 1973 and 1974.[13]

Hyams, the CBS reporter who had first imagined Apollo as a hoax, soon started work on a screenplay of such a narrative, finished in 1972. It told the story of the U.S. government

16.5
Robert Wussler in the CBS Control Room during
Apollo broadcasts.

secretly filming a Mars landing rather than admit a crucial failure in the spacecraft intended for the journey. Yet, as Hyams related to the *New York Times* years later, it was not until the House Watergate hearings in 1973 that the script was green-lighted by Warner Brothers. (It would become the 1978 Sam Waterston/O. J. Simpson vehicle *Capricorn One*.)[14] Explaining the delay, the *Times* concluded: "Watergate may not have inspired 'Capricorn One,' but it made its thesis more acceptable, its plot more credible and some of its content strangely prophetic."[15]

WHITEY ON THE MOON

After Apollo 11, lunar broadcasts saw further advances in broadcast technology. Starting with Apollo 14, a remarkable RCA camera rotated red, blue, and green filters in front of a single-channel camera to produce color images from the lunar surface. (Artifacts of the technology can be seen as colored streaks in rapid movement during broadcast, such as Apollo 14's Lunar Module take-off.) Yet color images did not increase audience share, nor did NASA's deliberate efforts during Apollo 16 and 17 to schedule the telegenic extravehicular activity, or EVA, for U.S. prime time. On the same album as his 1973 hit "The Revolution Will Not Be Televised," Gil Scott-Heron slyly raps the disconnect between everyday America and the image of Apollo; "A rat done bit my sister Nell / and Whitey's on the Moon."

As heralded by Scott-Heron, broadcast images of lunar exploration—with the exception of the suspense-ridden journey of Apollo 13—soon became a peripheral and even a discordant contrast to the tone of popular culture. Yet, as often as America's rapid drift away from images of lunar exploration is cited in contrast to the Kennedy-inspired commitment to the lunar landing, we would be succumbing to nostalgia if we did not see them as of a piece. Kennedy's understanding of the singular power of the image Apollo was designed to produce—the "dramatic achievement" he requested in 1961—was precisely that: an appreciation of a single image, not a long-running series. Instead of deploring the brief attention span of a media culture, we should marvel at the sustained attention paid to a single moment, and its continued reverberations in our own cultural firmament.

In the television saga of Apollo, a final note must touch on the central visual role played by the A7L spacesuit. The long CBS coverage of Apollo 11 began with a painted rendering of the image that was Apollo's true goal: For an entire minute, starting at 6:05 a.m. on July 16, the morning of Apollo 11's launch, viewers gazed into the open, golden eye of the A7L helmet, at the reflection not of their own faces but of the whole earth captured in the spacesuit's visor—from intimate to infinite, across the celestial void.

Notes

1. Jack Gould, "TV: Lunar Scenes Top Admirable Coverage," *New York Times*, July 22, 1969, 79.

2. See Gary R. Edgerton, *The Columbia History of American Television* (New York: Columbia University Press, 2007), especially chapter 6, "Television and the Presidency."

3. Erik Barnouw, *A History of Broadcasting in the United States*, vol. 3, *The Image Empire* (New York: Oxford University Press, 1966), 280–295.

4. Robert Caro, *The Years of Lyndon Johnson: The Path to Power* (New York: Knopf, 1982), 101.

5. David Halberstam, *The Powers That Be* (New York: Knopf, 1975), 438–439.

6. Ibid., 432.

7. Nicholas de Monchaux, interview with Mark Kramer, CBS Director of Special Events and head researcher, Apollo Broadcasts, New York, December 8, 2002. Audio recording. Also, interview with Joel Banow, CBS Apollo Broadcast Director, February 1, 2000, Joel Banow Collection, National Air and Space Museum Archives, Suitland, MD.

8. The transparencies in question were produced for NASA contract NAS-9-5981; NASA approved their use by CBS in November of 1968. See memo, Joel Banow file, NASM archives.

9. Interview with Mark Kramer, December 8, 2002.

10. Transcript of CBS Apollo broadcast, "10:56:20PM EDT 7/20/69: The historic conquest of the moon as reported to the American people by CBS News over the CBS Television Network" (New York: Columbia Broadcasting System, 1970), 67.

11. Benedict Nightingale, "What if a Moon Landing Were Faked? Ask Peter Hyams," *New York Times*, May 28, 1978, D10.

12. Halberstam, *The Powers That Be*, 655.

13. Making up for some of the gains of the executive branch in the previous decade, the Watergate hearings marked the first permanent entry of TV cameras into the House of Representatives.

14. Just as the Watergate conspiracy was unraveled by seemingly peripheral details, so too was the simulated conspiracy of *Capricorn One* uncovered by a small technical mismatch between mission telemetry and television signals, discovered by a Mission Control technician, and leaked to Elliot Gould's Woodward/Bernsteinesque journalist.

15. Nightingale, "What if a Moon Landing Were Faked?," D10.

Layer **17**　HARD SUIT 2

17.1

The AX-1 modeled by its custom-fitted occupant,
Vic Vykukal, 1964.

17.2

Vic Vykukal pantomiming a baseball throw in the
AX-2, 1967.

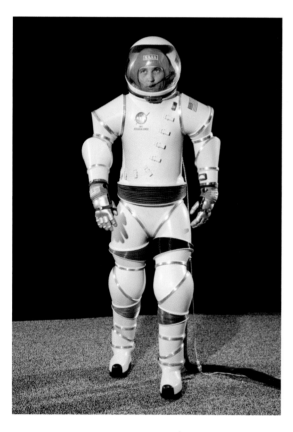

"I wouldn't go into space in something made on a sewing machine!"[1] So vows Hubert "Vic" Vykukal, lead designer of NASA's AX series of hard suits, built at the Ames Research Center from 1963 onward. There, on the outskirts of Mountain View, California, and in the shadow of the Navy's enormous dirigible hangar (see Layer 4), Vykukal spearheaded a series of prototypes that contested from first principles the development of extravehicular spacesuits from flight suit technologies. In his words, they attempted instead to "apply basic principles of engineering to the human body."[2]

On their own, and through transfers of their technology elsewhere, the AX suits provided essential competition to the soft suits developed for Apollo. In our narrative, they are also remarkable for the praise heaped on them by curators, designers, architects, and urbanists, and their prominent role in design literature.[3]

Unlike even the Litton RX suits—which had pleats of fabric behind their interlocking shells—the AX suits were entirely "hard," relying on a system of rotary bearings, tubes, and bellows to accommodate the range of human motion. The basic principles of such a concept had been explored for rigid diving suits as early as 1915, and, in anticipation of the need for extra-vehicular and lunar suits, were tested by Boeing and the Ames Research Center in 1963 and 1964.[4]

LOBSTERS AND LAYERING: THE AX-1

The spacesuit work at Ames developed from studies of hard, form-fitting acceleration couches in high-performance aircraft and early space capsules. Hand-cast to each pilot and astronaut, the fiberglass couches were expensive, and difficult to install and remove from aircraft and centrifuges. Starting in 1960, Ames engineers, led by Vykukal, developed a modular, hard acceleration couch that could be worn by the pilot and locked into standardized fittings in an aircraft or space vehicle. Such a device was experimentally fitted to an ILC pressure suit in 1965; a NASA press release billed its modular shells as a "Lobster Suit."

Working with the same fiberglass shell construction, the first full-body rigid spacesuit was manufactured in 1965: the AX-1. Made of hand-laid glass fiber, with bearings adapted from automotive drive shafts, and a suitcase-latch seal around the torso, the AX-1 was not intended as a finished product. Rather, it was an elegant proof of the possibility of a fabric-free pressure suit for use on the lunar surface.

While the waist and legs of the AX-1 used metal bellows to allow a degree of flexibility, the remaining freedom of movement provided by the suit came from rotational bearings joining oblique tubular sections: what Vykukal came to term a "stovepipe" joint. (In Vykukal's memory, the joint was inspired by an image of his father fitting a wood stove each winter into their two-room Texas home; the oblique joints embedded in the stovepipe allowed the custom-fitting of the chimney to its irregular roofline.) These techniques were combined to allow a remarkable degree of freedom to the suit's occupant. Today, even a slight jog to the AX-1 causes vibrations at every extremity, as the low-friction bearings translate forces through its entire surface.

COUTURE CARAPACES: THE AX-2

While the first, demonstrator AX-1 was custom-fitted—the addition of a quarter-inch clearance to a body cast of Vykukal provided the essential dimensions for the fiberglass mold—its follow-on, 1966's AX-2, was conceived as a fully scalable alternative to existing soft-suit designs. Its joints were slimmer and used more sophisticated bearings, in which new urethane plastics replaced the AX-1's Teflon lubricants. And between the joints of the suits lay a series of "sizing rings," cylinders of metal that could be added and subtracted between joints to fit a range of body dimensions. Added too was a series of indexing mechanisms to keep the many bearings of each joint synchronized to one another, and best conform to the range of human motion.

Like Litton's designs, and unlike soft suits, the AX-2 was a container, not clothing. Where the weight of a soft suit's leg, for instance, like that of earthly trousers, would be borne by the thigh and knee, the legs of the AX-2 were lifted from the sole of the foot. There, as on a racing bicycle, a clasp attached the fiberglass shell to the sole of a laced shoe worn inside. (A button was pressed on the suit's heel to release the laced foot from the shell.) Like a building, the suit had its own system of ducts and air vents laminated into the surface, incorporating into a single shell the multiple layers and tubes of competing soft designs.

The AX-2, despite its refinements, was never considered as an operational prototype. Many of its technologies, though, such as elegant shoulder joints, were quickly incorporated into commercial suit proposals. Indeed, when Litton Industries competed with the AirResearch division of the Garret Corporation (now Honeywell) for advanced Apollo suit designs, Vykukal consulted with each competing team in turn.

In the final competition for advanced Apollo suits (see Layer 15), AirResearch was victorious. As well as its own proprietary joint techniques, the AirResearch AES suit borrowed a variety of innovations from the AX suits. These included in particular a five-bearing shoulder joint, which had been designed by Vykukal to correct mobility limitations in the AX-2, and ensure that the suit fit in Apollo's narrow Command Module couches. Original innovations in the AirResearch suit included the use of laminated fabric panels instead of fiberglass for the suit joints; under pressure, the fabric laminates were as stiff as a hard shell, but at a fraction of the weight.

As we noted in Layer 15, however, neither AirResearch nor Litton suits would fly to the moon. A modified ILC A7L suit, the A7LB, was instituted for missions 14–17, and the remaining missions for which the AES suit was studied were canceled. A later proposal to use the AirResearch suit experimentally on the Skylab space station also came to naught, as NASA's by-then long experience with ILC suits overcame any claim by AirResearch to greater mobility and reliability in orbit.[5]

After the Apollo era, Vic Vykukal and the Ames Research Center continued occasional research into spacesuit design. A 1975 effort, the AX-3, explored a hybrid hard-soft suit, using laminated fabric layers, sealed to metal joints using high-powered magnetic pulses. The subsequent AX-4 was designed as a refinement of this research, but existed only as a set of drawn proposals sent to the Johnson Space Center in the late 1970s.

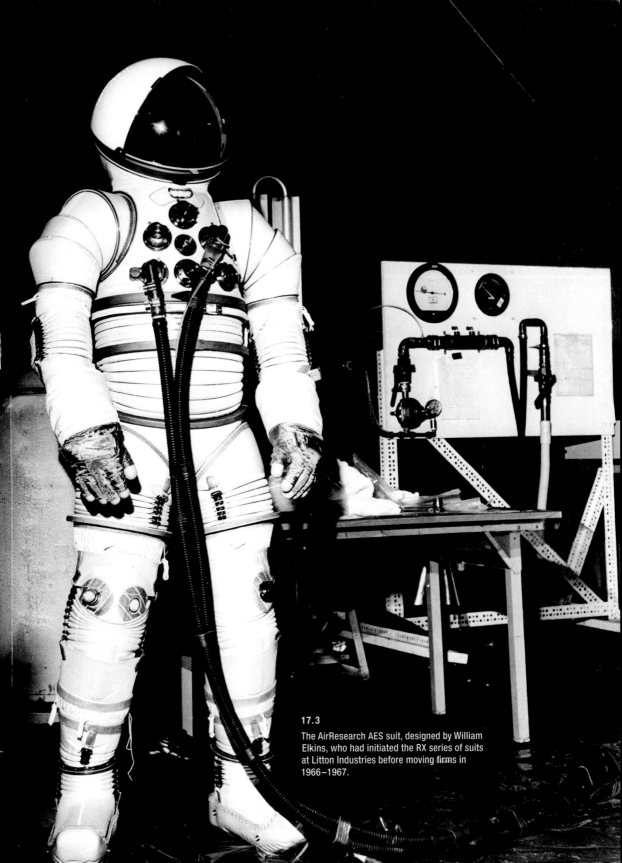

17.3
The AirResearch AES suit, designed by William
Elkins, who had initiated the RX series of suits
at Litton Industries before moving firms in
1966–1967.

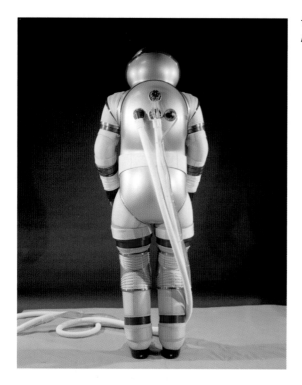

17.4
A rear view of the Ames AX-3, 1975.

SOLDIERS IN SPACE II: THE AX-5

The final suit in Ames's AX series, the AX-5, was not completed until 1988: long after the end of Apollo, and a decade after an ILC-fabricated suit had beat out competing proposals for the latest generation of Shuttle spacewear. It is nevertheless worthy of attention.

If Sputnik signaled the dawn of the spaceborne Cold War, and Apollo the conflict's apogee, then the AX-5 was born from its concluding skirmish. With the Reagan administration's launch of the Strategic Defense Initiative (known as SDI, or, popularly, "Star Wars") came the reconsideration of soldiers in space, a concept seemingly sidelined with the cancellation of MOL. Mirroring their predecessor's interest in the Litton RX-3, the 1980s Air Force was captivated by the durable surfaces and engineered clarity of hard-suit designs. But Reagan's Star Wars planners had even more pressing reasons to find an alternative to fabric suits.

Research into manned spaceflight for SDI did not focus on the spaceborne lasers or mass cannons that became the public image of the expensive proposal. Rather, it developed plans to use the Space Shuttle (for which a military launch pad at Vandenberg Air Base was under preparation until 1986), or even a top-secret military space plane, to launch missions that would echo the sabotage and spycraft of the Cold War below. Repairing, refitting, and sabotage of airborne space weapons were all considered as missions for Air Force astronauts, and each of these activities would expose a military astronaut to a range of hazards for which soft suits were considered specifically unsuitable.

At the most basic level, the polar orbits needed to access multiple satellites would involve repeated entry and exit from earth's Van Allen belts—dangerous fields of charged particles deliberately avoided by Apollo, and subsequent U.S. space missions. Operationally, manual access to operational satellites involved more tangible hazards as well. Corrosive rocket fuels, especially in Soviet satellites whose designs were not fully known, could quickly eat through the latex fingers of traditional suits. And there was the fact of a static-charged satellite as an Olympian doorknob, an arc from which might quickly melt through fragile, fabric layers.[6]

Spurred by the visit of Air Force General James Abrahamson to Ames in 1982, the AX-5 program spent tens of millions of dollars on a suit that could meet these operational parameters.[7] Weighing over 400 pounds, with a hollow torso milled from a one-ton ingot of aluminum, the AX-5 used the principles of the AX-1 and AX-2 to produce a heavily armored, operational spacesuit. The fiberglass joints of the AX-1 and 2 were transformed into bulky aluminum spheroids, precision-milled by Air Force contractors. The rotational bearings used in the suit benefited from military expertise as well, as similar devices were used in spy satellites; retooling the AX5's bearings to the military's exacting standards took three months alone.[8]

An entirely new problem addressed by the AX-5 had been avoided in all hard suits to date: the 36 degrees of mobility of the human hand, which all previous "hard" suits had accommodated with a soft, latex-and-fabric glove. Given the hazards expected in military spacewalks, however, a hard substitute was necessary. The AX-5 pioneered such a glove using the bearing-and-bellows strategy developed since 1964. The gloves can be observed in photographs of the AX-5's underwater testing—the suit was too heavy to be used outside of a water-tank's buoyancy (or, it was hoped, in the weightlessness of orbit).

17.5
The AX-5 in underwater testing, 1983.
Vic Vykukal is in the suit; Ames suit engineer
Bruce Webbon is at the tank window.
Note hard gloves.

"THE BODY AUGMENTED AND HOUSED"

"It's a beautiful suit," commented NASA Crew Systems engineer Matt Radnofsky in 1964, on first viewing the AX-1 prototype.[9] An enduring legacy of the AX series of suits has been their aesthetic appreciation. Not only were NASA officials—like Radnofsky—swayed by the suits' purity of line, but so, eventually, were the design professions themselves. Unique among pressure suit assemblies, the AX series of prototypes has occupied pride of place in design criticism, particularly in the body-conscious 1980s.

In 1982, the Ames AX-3 appeared, along with the Bauhaus designs of Oskar Schlemmer, and Soviet constructivist costumes of Nadezhda Lamanova and Alexandra Exter, in the catalog of a contemporary fashion exhibit, "Intimate Architecture," at MIT's List Visual Art Center in Cambridge, Massachusetts. While a stock photograph of the AX-3 was used, more contemporary fashions—geometric constructions by Miyake, Ferre, and Armani—were lushly photographed by Robert Mapplethorpe. Comparing the AX-3 directly to constructivist speculations, curator Susan Sidlauskas writes: "If a more comprehensive response to a hostile environment is required, the elegantly shaped, self-sufficient NASA AX-3 space suit may suggest both the aesthetic and the practical solution."[10]

A decade later, scholar Georges Teyssot featured the AX-3 in an introductory essay to the monograph *Flesh*, by architects Diller + Scofidio, whose mechanically inflected intimate architecture included costumes, gallery settings, and designs for domestic spaces. Set alongside works by Cristobal Balenciaga and the electronic prostheses of performance artist Stelarc, the AX-3 is admired as "simultaneously essential and superficial."[11]

And in 1987, in an essay that featured the early work of Diller and Scofidio with that of Coop Himmelbau and John Hejduk, critic and urbanist Michael Sorkin deemed Vic Vykukal "one of America's great hidden design geniuses." Sorkin dubbed the AX-5, in particular, "one of the most beautiful designed objects I've ever seen, at once sublimely functionalist and wackily Schlemmeresque."

"It is also," Sorkin continues, "an apparatus of liberation, of extension, of genuine prosthesis, the body simultaneously augmented and housed."[12] Recalling Buckminster Fuller disciple John McHale's 1967 *Architectural Design* essay "Man +"[13] (see Layer 19), the supremely engineered, rigid AX-5 is presented as an optimal vehicle for the extension of, and integration with, the human body. Heralding Sorkin's later zeal, McHale wrote in 1967: "The record of technological development is one of a progressive overlay of another form of evolution on the natural genetic process." Harking back further to the modernist rhetoric of the Bauhaus, or the work of Soviet constructivists with which the AX suits would be compared, Sorkin's rhetoric extends the utopian technological promise at the heart of twentieth-century modernism.

This ideal, in many ways, was at the heart of the AX suits' inception as well. Recalling his own inspiration for the suits, Vic Vykukal questions the basic assumptions of fabric pressure suits: "Dirigibles and biplanes were fabric; could you fly *them* supersonically? And cars haven't had inner tubes for 50 years. So why are spacesuits made of fabric and inner tubes?" To many NASA engineers, soft pressure suits were "junk," not simply because they were an early form of suit development, but because they relied on kinds of handwork, sewing in particular, that were antithetical to the normal methods of military-industrial manufacturing.[14]

Yet we should perhaps soften such a harsh conclusion. We have seen elsewhere how difficult was the interface between ILC and NASA's aerospace culture; with sometimes literal interfaces to weaponry and war, the armored shells of the AX and RX suits were fashioned deliberately, and suitably, for their own organizational setting.

With the exception of the AX-5, we also see in the subsequent "hard" Apollo prototypes a softening as well. While still eschewing a sewing needle, the later, harder competitors to ILC became increasingly pliable. Whether Litton's Constant Volume Suit—an AX-inspired revision of the RX designs—or the AirResearch EX1 and AES (masterminded by William Elkins after his departure from Litton), the suits abandoned fiberglass for inflatable fabrics. And while these fabric suits appropriated more traditional aerospace methods—they were "bonded," not sewn— the "hard" suits that came closest to flying on Apollo were, in their own way, increasingly "soft" as well.

Notes

1. Nicholas de Monchaux, interview with Hubert (Vic) Vykukal, November 1, 2007, audio recording.

2. Vykukal interview.

3. Michael Sorkin, "Minimums," *Village Voice*, October 13, 1987, 100.

4. Gary L. Harris, *The Origins and Technology of the Advanced Extravehicular Spacesuit*, American Astronautical Society History Series, 24 (San Diego: Univelt, 2001), 186–187.

5. Nicholas de Monchaux, interview with William Elkins, June 2, 2009, audio recording. See also Harris, *The Origins and Technology of the Advanced Extravehicular Spacesuit*, 183.

6. Vykukal interview, November 1, 2007; also Harris, *The Origins and Technology of the Advanced Extravehicular Spacesuit*, 354–355.

7. Vykukal interview, November 1, 2007; Harris, *The Origins and Technology of the Advanced Extravehicular Spacesuit*, 354–355; Kenneth S. Thomas and Harold J. McMann, *US Spacesuits* (New York: Springer Verlag, 2006), 305–307.

8. Vykukal interview, November 1, 2007.

9. Ibid.

10. Susan Sidlauskas, catalog to the exhibit "Intimate Architecture: Contemporary Clothing Design," Hayden Gallery, Massachusetts Institute of Technology, Cambridge, MA, May 15–June 27, 1982, introductory essay (unpaginated).

11. Georges Teyssot, "The Mutant Body of Architecture," from Elizabeth Diller and Ricardo Scofidio, *Flesh* (New York: Princeton Architectural Press, 1996), 12.

12. Sorkin, "Minimums."

13. John McHale, "Man +," *Architectural Design* 37 (February 1967): 85–89.

14. Vykukal interview, November 1, 2007.

Layer **18**   CONTROL SPACE

In Layer 12 we examined the origins and legacy of a certain kind of nonarchitecture—the virtual spaces used for training pilots and astronauts, up to and during Apollo. In the current layer we take a related detour through the enormous looking glass of the space race's control rooms. Like the virtual spaces of Whirlwind, the multiscreen control spaces of midcentury had their beginnings in America's postwar nuclear defense. Developed in symbiosis with the virtual spaces of aircraft and spacecraft simulators, these multiscreen theaters would have their public apogee in the stage-set of Apollo's Mission Control—a two-word phrase inseparable from the architecture of the space race, and, it turns out, having its own claims on the human body as well.

STAGES, VIRTUAL AND REAL

While the virtual space of SAGE's nuclear defense extended across land and ocean, SAGE control rooms were made of standard-issue office flooring panels, their white surfaces similar to the spotless computer rooms that IBM and others would deploy for years to come (IBM, indeed, was the main supplier of SAGE equipment).[1] The promise of real-time control imbedded in the glowing phosphors of the SAGE CRT, however, would soon seed the ground of another architecture of control and simulation. Much larger in scale, it would in many ways be equally immaterial.

The basic DNA of control rooms existed long prior to SAGE, in the "combat centers" and other "situation rooms" that had come to play a particularly central role in WWII decision-making.[2] Sometimes static, often mobile (as on an aircraft carrier), these were singular spaces in which information about forces was posted for the aid of commanders. The first physical attempt to update such spaces for nuclear realities came in a meeting room constructed for the newly independent Air Force's Directorate of Intelligence, where it began to strategize responses to the newly perceived threat from the Soviet Union.

A prototypically immaterial information space, the postwar war-room had no walls. Its conference table was ensconced within curtains, each curtain hiding sliding panels of information about Soviet troop levels, positions, and foreseen capabilities, manually reconfigured by officers in adjacent rooms. This was a space of command, but not necessarily of mechanical control; information flowed into the room via flexible panels and signage, but commands were released from it verbally or by telephone. The next step in the development of this particular Cold War typology would extend this model to create a space of display and control for the Strategic Air Command (SAC) at Offutt Air Force Base in Nebraska.

18.1
Air Force Intelligence Directorate briefing room;
view showing detail of sliding information
board mechanism.

Unlike the Air Force Directorate of Intelligence sitting in the Pentagon, SAC commanders were physically dispersed. To broadcast spatial and numerical data to many different locations, the Command built a giant soundstage under the Nebraska prairie, an armored TV studio whose enormous rear wall contained the same kinds of charts and maps used in more conventional presentations. Updated constantly by ladder-borne clerks, the boards' information was transmitted from the command post through banks of television cameras on the rear wall along secure TV cables. By the 1960s, the architecture of the underground bunker at Offutt was altered to accommodate more dynamic displays as well as the consolidation of command post functions in a viewing gallery. New transparencies for these screens could be prepared in as little as 30 seconds.[3] In a later, classified photo from the 1950s, banks of dedicated terminals have replaced office desks on the soundstage floor, and one side of the room features a projected CRT display of U.S. airspace. The room has become not just a soundstage for display, but also a stage for action.

18.2

The Offutt Air Force Base soundstage in a declassified 1959 photograph.

18.3

A later view of the soundstage showing overhead projection, and a dedicated series of display terminals, that would foreshadow the parallel architecture of Mission Control, undated, declassified 1963.

18.4

General view of the Mercury control room during the flight of Walter Schirra and the Mercury 8 spacecraft.

18.5

A 1963 rendering of the Mission Control Center in Houston, one of two identical rooms planned for the complex.

NASA

Meanwhile, the public face of the Cold War, the space race, was creating its own control spaces. Developed at the same time as SAC's bunkers, and sharing both layout and equipment with them, these "Mission Control" rooms had their own roots in the concrete blockhouses dug into the sands of Cape Canaveral, control rooms for missile tests that had started with Bernard Schriever's ICBM program in the 1950s.

The first of these were cramped, airless affairs related to the mobile control vans that had launched von Braun's first V-2 rockets from the streets of Antwerp and The Hague in the Second World War. Then, as in the sands of Florida, such rooms were dependent on instantaneous communication with the rocket and the infrastructure of its launch. By the launch of the first Mercury-Redstone rockets (themselves modified V2s), the Cape Canaveral blockhouse was a forest of indicator lights and small TV screens, each human operator in the launch command assigned a fixed position among the controls and display tubes.

As rocket missions started to last much longer than a launch test, NASA created a new kind of space adjacent to the Cape Canaveral blockhouse, termed the "Mission Control Center." The room was a hybrid of SAC's Nebraska soundstage and the launch control center located nearby. One end of the room was taken up with a display surface showing the mission trajectory. Like SAC's soundstage, as well as its World War II ancestors, information on the control room wall was updated by hand. In the new Mission Control, however, technicians became stagehands, shielded from view as they continually updated screens of data. In front of the display wall, dedicated control desks, wired to the systems and computers on the ground and in communication with those aboard the orbiting capsule, populated the room. The only truly instantaneous visual information, however, was provided by three television sets attached to the ceiling. By the time of the later Mercury missions, modifications had been made to the control room; instead of manually updated charts, a portion of the front wall was replaced with an early TV projector showing computer-generated information.

As characterized by a NASA Langley official considering the requirements for a Mission Control Center to replace the Mercury Control Room, the subsequent Gemini and Apollo missions of the lunar plan were "*nothing* [original emphasis] like Mercury." Gemini and Apollo—an uncertain trajectory of missions in which astronauts would spend weeks in space learning how to dock with other spacecraft, testing endurance and technological constraints before progressing to lunar orbit, rendezvous, and landing—generated a correspondingly complex control space. By 1961, over 30 million dollars was budgeted for the new Mission Control Center (MCC) in Houston.

Well-qualified by its work on SAGE, IBM was contracted in 1962 to build "information systems" for the new facility. By 1963, they had been joined by Philco-Ford, which would provide the "Display and Control System," the visual interface between Mission Control's human operators and computers on board spacecraft and on the ground. The innovations wrought by IBM in building Mission Control's computing systems would ultimately flow into the firm's larger product lines.[4] Philco-Ford's work on displaying that information was, by contrast, a more specialized affair.

A full description of Mission Control's 1960s display technology would run to many volumes of technical reports. The system was an enormous, and enormously expensive, hybrid of the era's

18.6

One of two identical Mission Control Rooms opened
in 1964. The details of a 1966 Gemini flight are
visible onscreen.

state of the art. Instead of the transparencies or manually adjusted charts of previous control rooms, the main display surface of Houston's MCC had a dozen computer-controlled projectors. For the largest displays, which showed maps and orbital trajectories, color slides of the earth (and moon) were overlaid with linear plots and text scribed dynamically by diamond-tipped styluses into lead-coated slides. High-intensity krypton lamps of various colors were projected through the scribed openings to create multiple-layer displays of spacecraft telemetry. Even more complex were the television projection mechanisms adjoining the map surfaces on the large display wall. These enormous Eidophor projectors used platters of electrically charged oil to create three large TV screens on the rear wall. The 48 channels shown by these screens were separately accessible at consoles on the control room floor. Each channel was constructed from its own combination of color slide transparencies and vector-based "Charatron" digital projection, and transferred through a system of mirrors and high-resolution TV cameras into a computer-controlled "Video Switch Matrix."[5]

In front of the huge screens, individual buttons on each control desk were reconfigured for each separate Gemini and Apollo mission. In the Gemini era, this was done with a punchboard; only a few years later digital switches performed the same function instantaneously. In an echt-steampunk symphony, the push of a single button on each console triggered a photographic system adjacent to the video matrix, which recorded the television image being consulted by the controller, and sent a rapidly developed 8-by-10-inch image of the screen via pneumatic tube to the controller's desk within 60 seconds.[6] The system comprised more than 125 control consoles, 400 television screens, 130 television cameras, 20,000 video switches, and 84,000 35mm background slides.

Whether real or illusory, the sensation of control conveyed by the vast Mission Control Complex was a hyper-sensory synthesis of image and text that recalls contemporaneous experiments in cinema, such as Charles and Ray Eames's 1959 "Glimpses of the USA." The twelve-minute, seven-screen spectacular was exhibited in Moscow in 1959, and, it is claimed, took its own inspiration from early control room renderings.[7] In the Eames's staging, thousands of still images were sequenced on 35mm film projected onto separate 20-by-30-foot screens. The theater containing the show was the largest geodesic dome constructed by Buckminster Fuller to date, spanning 250 feet and housing 5,000 viewers 16 times a day.

A more enduring Eames architectural legacy is the organizational design of midcentury, considered by architectural historian Reinhold Martin in *The Organizational Complex: Architecture, Media, and Corporate Space*. Martin compares the mute exterior of the IBM 360 (three of which formed part of Mission Control's systems as of 1965) with the mirrored, visually elusive exterior of the corporate buildings in which the IBM computers were often housed. Mission Control, whose interior was far more phantasmagoric, was even more camouflaged—a windowless white building set against parking lots and a well-tended lawn. Like most of Houston, the Mission Control complex (now part of the Johnson Space Center) had been built from scratch on marshy farmland. Visitors to the complex often needed directions to find Mission Control among an identical sea of service buildings likened by one critic to "a series of ophthalmologist's offices at a suburban mall."

18.7

Manned Spacecraft Center Building 39. Mission Control rooms are housed in the largest windowless structure on the left-hand side of the image.

18.8

Seismically isolated NORAD computer rooms under construction under Cheyenne Mountain, 1964.

If part of the nature of Mission Control was a facadeless interior, then the same quality can be found in extremis in Mission Control's secret sibling. At the same time as it was constructing the visual display systems for the anonymous Building 30 in Houston, Philco-Ford was instituting a near-identical system deep inside the bowels of Cheyenne Mountain, Wyoming.[8] In response to the threat of ballistic missiles, the mountain would house a new North American Air Defense Command (commonly known by the acronym NORAD), which would amalgamate functions from SAC and the Air Force war-room in Washington.

Supposedly resistant to a 1.5 megaton nuclear attack, the mountain labyrinth would grow to 4.5 acres of caverns, inside of which were constructed steel frame buildings isolated by rubber bearings from the mountain rock, up to three stories tall. A photograph of these buildings under construction confounds our architectural sensibilities. Modern, Miesian, and corporate, they lie within a Brothers Grimm cavern, sealed not (like Mission Control) by suburban banality but rather by a more epic violence—the Farnsworth House in funeral rites. If Houston's Mission Control had no exterior to speak of, then NORAD would have no facade at all, a brain without a face.

SIMULATION

The ironic reality of computing history is that the more responsive and immediate the illusion of control created by steadily computerized military-industrial control rooms, the more the rooms were used to operate on imaginary conditions instead of real ones. Indeed, the conditions that these spaces were dedicated to—whether lunar landings or thermonuclear war—were, as of 1964, equally fantastic. The central goal of such spaces was to create the circumstances and procedures whereby the unimaginable could ostensibly be controlled and practiced for, so that when it arrived in reality (the hoped-for lunar landing on the one hand, the feared-for Armageddon on the other) it could be expertly managed thanks to experience gained through simulation. Control rooms were built like theaters because they were theaters—confined spaces by which the fantastic can be examined, understood, and incorporated into the fabric of reality. NASA termed their extensive simulations "full dress rehearsals."[9]

Houston's Mission Control had one sibling entombed in Cheyenne Mountain, but, to allow such extensive simulation, it also had a Siamese twin. By 1965 a conjoined, identical control space had been stacked on top of the first Mission Control room. Even as the lower control room took over Gemini missions, the upper chamber was used to simulate upcoming Apollo missions. With the end of Gemini flights, the two rooms were devoted to Apollo, with upcoming flights simulated atop and "real" missions underneath as they progressed.[10] In his 1967 *Architectural Design* essay, "Man +," John McHale used side-by-side illustrations of NORAD and Mission Control to symbolize the power of systems thinking and its incorporation in to human affairs. They were evidence, he claimed, of Marshall McLuhan's sweeping assertion: "today … we have extended our central nervous system itself in a global embrace."[11]

CURRENT TRENDS IN COMPUTER ARCHITECTURE

In a 1959 essay about a new machine developed by IBM,[12] engineers F. P. Brooks, Jr., G. A. Blaauw, and Werner Buchholz were the first to use the word "architecture" to apply to the interior arrangement of computers. The device in question, the IBM 7030, or "Stretch," proposed a re-arrangement of the computer's interior circulation of information to achieve greater usefulness and functionality; to the engineers, this was in essence identical to the design of physical space toward the same ends.

In this context, and with its specially adapted IBM mainframes, Mission Control was a recombinant architecture of information, astronaut, and technician. It was not just Mission Control and simulation that were moved from separate sites to Houston in 1965, but the staff of programmers dedicated to NASA's information systems as well. Advocating such a move in 1962, NASA's then Virginia-based Flight Operations division engaged in a mock-dialog as part of a 20-page memorandum—holding out the prospect of complete failure if the colocation were not approved.

> Do the programmers *really* have to work directly with the mission designers and flight controllers?
>
> Absolutely yes! Regardless of who they are! They are doing a single task. Mission design logic and computer logic is one and the same. It is inseparable. That is the entire point. Mission designers and computer programmers must evolve this complex logic together so that each has a complete reciprocal understanding of what the other needs and what the other is doing in detail.
>
> What will happen if they are not located physically together?
>
> Chaos! Literally chaos![13]

The resulting collaboration of operators and designers of procedures, and operators and designers of information, had consequences in worlds virtual and real. Indeed, such blurring has come to define contemporary experience.

Within the world of computing, NASA's control systems had physical consequences—the first solid-state computer using silicon wafers ever built was the Apollo guidance computer. Equally important, however, were the effects of NASA's systems on computing's more abstract architecture, that of information flow.

As already noted, the notion of "computer architecture" grew from the development of the IBM 7030, and its then-unique data-processing structure. Instead of one or two dedicated and differentiated "registers" for the storage of data operated on by the computer, the 7030 literally "stretched" the number and type of registers to 16 so-called "rooms" in which sets of data could be stored and transformed by each other. Instead of retaining the batch processing of earlier IBM computers—itself a legacy from literal batches of punch cards that would be lined up and processed one at a time—the 7030 could process data in real time, and change and alter its priorities based on circumstances. While the 7030 was a commercial failure for IBM, this architectural shift would be seismic, allowing computers to transform from calculators supplying data for decisions to decision-making devices operating in real time, and in partnership with their human operators. With the incorporation of the 7030's innovations into Apollo's hardware and software, the first citizens of the space age were pioneers of this new information age as well. The 7030's innovations were directly introduced into IBM's specially adapted Mission Control mainframes. Thus, NASA's information architecture was transformed from a sequential enfilade to a linked network of information spaces that resembled nothing so much as the cellular warren of Mission Control's own Building 30. The most important cell of this network was Building 30's room 112, the "Real-Time Computer Complex" (RTCC). There, five specially adapted IBM 7090II computers had been modified from their original sequential processing to this new, network architecture, simultaneously addressing "all computable data supporting actual and simulated missions."[14]

A 1969 article in *Time* explained the dynamics of Mission Control and emphasized the link between the physical architecture of the room and the digital architecture of the computers adjacent to it:

> "I'm like an orchestra conductor," says Christopher Columbus Kraft, flight operations director for the Apollo missions. "I don't write the music, I just make sure it comes out right." Chris Kraft's unlikely podium is the windowless Mission Operations Control Room on the third floor of Building 30 at NASA's Manned Spacecraft Center near Houston. His musicians are the 30 controllers who sit at four rows of gray computer consoles, monitoring some 1,500 constantly changing items of information registered on gauges, dials and meters. Kraft's primary instrument is a pair of IBM 360 Model 75 computers with a total capacity of 2.5 million bits of information, which enables him to harmonize the thousands of complex equations and manifold instructions that program a lunar mission.[15]

In 1962, when the five computers of the RTCC were first installed, the software that allowed these computers to prioritize the data flowing into and out of their processors, based on priority to the mission, was called "Mercury Monitor," after its first deployment in Cape Canaveral's Mission Control Center. By 1966, when the five 7090II computers in Room 112 were replaced with new IBM system 360 computers (the new model's architecture derived directly from "Stretch"), the system for prioritizing data was known as the Houston Automatic Spooling Priority, or HASP.

HASP would have its own, enormous, legacy. Originally written exclusively for NASA to better exploit the System 360's lattice-like architecture, it became by the mid-1970s an enormously successful IBM product in its own right. HASP was the best, and in many ways the only, operating system that could adapt to new control systems in transportation, government, and finance, administering checking accounts, interbank transfers, and airline reservations nationwide. Until the dawn of the personal computer a decade later, it would be IBM's best-selling software.[16]

INDIVIDUALISM SUBSUMED
In his biography of Buckminster Fuller, Lloyd Sieden argues that the famous systems architect's world view received an important early influence from his study at the Naval Academy in 1918.[17] There, in an era before reliable radio communication with ocean vessels, Fuller received a comprehensive education in every system of a ship at sea he might need to engage, from boiler room to bridge, up to and including the medical treatment of his crew.

18.9
Flight controller, Mission Control, 1965.

18.10
Mission Control during the Apollo 11 lunar EVA.
Note coffee and cigarettes.

18.11

From left to right, Aldrin (obscured), Collins,
and Armstrong walk along a corridor at
the Kennedy Space Center on their way to the
Saturn V/Apollo 11 vehicle.

For a time in the planning of Apollo, a similar strategy was contemplated. A "map and data viewer" in the Apollo capsule was to contain "thousands of different frames of information, very high information density ... [a] combination road-map, almanac, and manual of special instructions."[18] Proposed in 1963, the device was never implemented; control from the ground of the Apollo spacecraft increasingly won out over those arguing for the astronaut's autonomy. In a small victory for the astronauts, director of the Apollo Astronaut Office Deke Slayton successfully argued against a teleprinter in the Apollo capsule to send written instructions to the crew. As a larger trend, however, the space age subjected astronauts to what historian Walter A. McDougall has described as an "individualism ... subsumed into the rationality of systems."[19]

A photograph of Armstrong, Aldrin, and Collins in 1969 depicts the trio walking down a hallway in the Kennedy Space Center on their way to the Saturn V Rocket and Command Module at its summit. Encased in their suits and carrying portable air-conditioning equipment, the astronauts seem not to be heading into space but, rather, to have arrived on earth as aliens, separated and set apart from the banal background of ceiling tiles and computer consoles by the layered fabric of their spectacular garb. Dressed already for departure from the earth, they seem to belong anywhere other than the clean-room corridor. As perceived by Kubrick in *2001*, however, clean-room architecture—home not just to preflight astronauts but also to the systems and computers that controlled them—was as much, or more, the architecture of the space race than the clinging dust of the lunar landscape or the inhospitable vacuum in which it is set.

As much as the photograph sets up a poetic tension of arrival and departure, and as much as the astronauts are defined by their incongruity in the landscape of organization and information, it is most important to understand that this landscape—the bright banality of fluorescent lights and raised-panel flooring—was one they would never leave, even a quarter-million miles distant.

CONTROL ROOMS VERSUS CUSTOM-FITTED FLEXIBILITY

The April 1975 *Apollo Program Summary Report*, NASA's 800-page summation of the 15-year Apollo effort, contains only a few paragraphs on the A7L suit. It is described, in passing, as a "multilayered, custom fitted, flexible garment."[20] In its flexibility and customization, the A7L was nevertheless central: one of the only designed interfaces between the biological, organizational, and psychological complexities of the human body, and the command infrastructure of the larger Apollo effort. But other such interfaces existed; they would become visible especially at moments of great tension between the highly regimented technologies and procedures of Apollo's military-industrial DNA and the inescapable, unpredictable reality of human life.

LUCKY 13

02 07 55 20 CDR: I believe we've had a problem here.

02 07 55 28 CC: This is Houston. Say again, please.

02 07 55 35 CDR: Houston, we've had a problem.

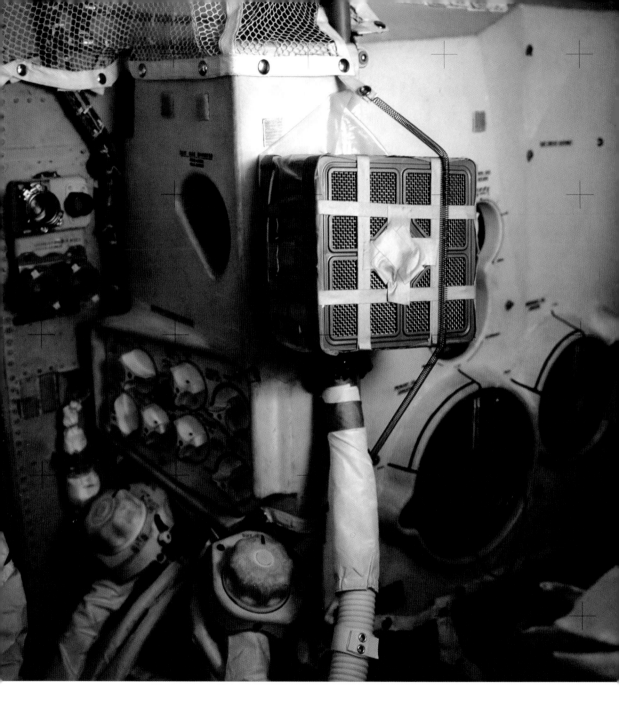

18.12
The improvised air filter constructed by the
Apollo 13 crew. Note duct tape, fabric, and
spacesuit hoses and fittings.

This exchange, two days, seven hours, and 55 minutes into the flight of Apollo 13 toward the moon, took place between mission commander James A. Lovell and capsule communicator Jack Lousma, and marks the moment when the spacecraft's long-planned trajectory went perilously awry. As the result of both a physical defect in an oxygen tank and the much larger administrative breakdown in systems engineering that both allowed and failed to detect the fault, a massive explosion ripped through the mission's Service Module. The ship was desperately crippled, and its crew nearly stranded in interplanetary space. The drama that followed, with an improvised, duct-tape-wrapped oxygen filter allowing the three-man crew to use the two-man Lunar Excursion Module as a lifeboat for the return trip to earth, is often celebrated. Yet that arduous return to earth was not without its frictions.

> 04 02 26 48 LMP Jim doesn't have his BIOMED rigged on right now …
>
> 04 02 27 07 CC Okay, Fred. You're too weak on that last. I understand you're disconnecting, and your BIOMED will be off. Say again about Jim, please.
>
> 04 02 27 24 LMP Okay. How do you read now?
>
> 04 02 27 26 CC Okay.
>
> 04 02 27 30 LMP Okay. All I said was Jim doesn't have his BIOMED rigged right now …

Hidden in this second transcript is a rebellion of sorts on the part of Jim Lovell, Apollo 13's Mission Commander. Violating strict rules and coping with an unprecedented amount of stress and uncertainty, Lovell had removed the electrodes that had been glued to his body since before launch. These relayed continuous biomedical data to Houston, where they formed part of the matrix of data accessible to Mission Control. In his recounting of the Apollo 13 drama, Lovell enumerates three reasons for his actions: first, the sensors itched, and second, removing them would save a small amount of power in the stricken craft. Finally, however, Lovell cites privacy. Speaking in the third person, he writes: "Lovell did not know how high his cardiac rate climbed after the explosion that aborted his mission … but it rankled him to know that everyone from the flight surgeon to the FIDOs [flight dynamics officers] to the pool reporters did."[21] In severing the electronic link between his own heartbeat and the control room far below, Lovell was very directly asserting the autonomy of his own body from the very technological and institutional forces that had placed it so spectacularly in jeopardy.

If Lovell's electrode removal was a barely observed marking of the line between man and machine in Apollo, a far more dramatic rebellion took place during the penultimate flight of Apollo hardware 1973–1974's Skylab IV. During this, the third manned sojourn on the short-lived space station, an unprecedented series of conflicts occurred between the massive staff of Mission Control and the three astronauts, Gerald Carr, William Pogue, and Edward Gibson.

Problems on the 84-day mission began only a few hours after the astronauts' entry into the station. William Pogue, feeling disoriented from the ascent to Skylab and the immediate burst of activity on entering the large, open station, threw up. While mission regulations decreed that the event should be reported—and the vomit freeze-dried and brought back to earth for analysis—the astronauts did neither. While secreting the evidence in the station's trash repository, they remarked to each other:

Carr: We won't mention the barf, we'll just throw that down the airlock.

Gibson: They're not going to be able to keep track of that.

Pogue: Yeah, it's just between you, me and the couch.

All the while forgetting that, in the words of New Yorker writer Henry S. F. Cooper, "Skylab was bugged as thoroughly as the Nixon White House."[22] A transcript of the remarks was widely distributed throughout NASA's bureaucracy the next day, and resulted in a strong reprimand being relayed through capsule communicator Alan Shepard.

Successfully resisted in Apollo's earlier incarnations, an instructional teleprinter was installed at the entrance of Skylab's crew compartment. The astronauts' day was described by flight director Neal Hutchinson: "We send up about six feet of instructions to the astronauts' teleprinter in the docking adapter every day—at least 42 separate instructions. ... We lay out the whole day for them and the astronauts normally follow it to a T. What we've done is, we've learned how to maximize what you can get out of a man in one day."[23]

The crew's response to the punishing schedule of instruction was ultimately dramatic. On December 12, in the fourth week of the mission, the Reuters wire service reported on "the slowness of the crew and the lack of enthusiasm."[24] By December 27, the situation had degraded to the point where, after an angry diatribe from orbit, the crew took an unplanned day of rest, and (in circumstances that are still contested) switched off all communication with Mission Control for several hours. In the words of Commander Gerald Carr: "we took our day off and did what we wanted to do."[25] A Harvard Business School study termed it the "Strike in Space."[26]

As well as bemoaning their heavily programmed, 16-hour daily schedule, the astronauts of Skylab took particular exception to the more intimate expressions of systems management. Of Skylab's interior design, Owen P. Garriott, the otherwise even-tempered mission scientist of the previous Skylab crew, complained: "It seems to me that the color arrangement that we've got in here might very well have been designed by a Navy supply department or something with about as little imagination as anybody I can imagine! All we've got in here are about two tones of brown."[27]

"I just get tired of this darn brown!" Skylab IV scientist Edward Gibson lamented—not about the Skylab interior, but about his own clothing, supplied only in a yellowish-brown fireproof polyester. And of the (also brown) fireproof towels used for washing,[28] Skylab IV mission pilot William Pogue judged that using them was "sort of like drying off with padded steel wool."[29] Throughout the Skylab missions, astronauts worked to modify their clothing and sleeping bags to make them more comfortable. Alan Bean, commander of the prior Skylab III mission, particularly lamented the failure to include a needle and thread among the station's 40,000 items of cargo.[30]

18.13

Astronauts Carr and Pogue squeezing trash bags into Skylab's airlock; the empty fuel tank on the orbiter used for trash collection was quite full by the end of the final mission.

The physical appearance of the Skylab IV crew, combined with their "blistering language," made Houston "uneasy," especially their "thick, revolutionary-looking beards."[31]

When reporting the results of their mission to Congress, the Skylab IV astronauts made the following statements:

> Dr. Gibson: ... systems have to be designed and missions in particular have to be designed so that you have ways in which the human can continually use his judgment ... we would like to see future missions make more use of this unique capability of man ...

> Colonel Carr: Our key point is if you are going to use man in space he is going to be happy if he can use his judgment and if he can feel he is productive. If he is nothing but a switch-twiddler working against a clock, he is going to become very bored, and he is going to have psychological problems.[32]

The astronauts said little else about the personal dramas of their mission in the congressional briefing. Remarkable, however, is the emphasis they placed on the effects of mankind on earth, instead of the natural and astronomical phenomena they were ostensibly meant to study. Carr, Gibson, and Pogue all remarked on their study of urban growth and form from space, and, most extensively, their orbital examination of air pollution, strip mining, wetland conservation, clear cutting, and the dumping of chemicals into lakes and oceans.[33] Of the 26 slides shown by the group of astronauts to Congress, 17 depicted ecological damage wreaked by mankind.[34] Unfortunately, the lack of public interest in the final manned missions of the 1970s muted the crew's environmental message. Only a few congressmen on the House Science and Astronautics subcommittee actually attended the hearing in question, and the Skylab IV crew's splashdown and recovery was the first Apollo landing not to receive live television coverage.

In summing up the effects of the flight on his personal outlook, Colonel William Pogue reflected: "I now have a new orientation of almost a spiritual nature. ... When I see people I try to see them as operating human beings and try to fit myself into a human situation instead of trying to operate like a machine."[35]

CONCLUSION

The architecture, even the notion, of "Mission Control" is essential not just to the story of Apollo, but to the twentieth century itself. Seeded by the time-and-motion studies of the assembly line, and translated through operations research into systems engineering, the multiscreen control space appeared first as a weapon of war—to optimize the split-second, life-and-death decisions of the Strategic Air Command. When the Philco-Ford screens of NORAD became Houston's Mission Control, however, we are faced with a prospect that literally recalls the Clynes-Kline cyborg—physiological variables nested with, and inseparable from, orbital mechanics, fuel loads trajectories, and the other native languages of the military-industrial complex. As well as blurring the boundaries between man and machine, the systems architecture of NASA also worked to erase any subjective distinction between reality and the new, virtual spaces of the space age. Apollo technicians commented that it was almost impossible to tell which of the twin Mission Control rooms in Building 30 was running a "real" mission; the inputs, outputs, and displays were identical in either case. Not only did the training for Apollo's lunar landing pioneer the computer-mediated experience of virtual space, the actual landing relied on it. Within the systems architecture of the Apollo Guidance Computer, as well as its earthbound simulation in Building 30, lived a tiny, virtual moonscape, the lunar terrain model, used to calibrate the range-finding routines leading to touchdown.[36]

In the context of our narrative, and the literal mediation of the "multilayered, custom-fitted" spacesuit between biology and technology, the gaps and fissures in the multiscreen, Mission Control world view are of particular importance. As engineer and historian David Mindell empha-sizes in his history *Digital Apollo*, the elaborate design of an automatic landing system for the Lunar Excursion Module was never used; in five successful lunar landings, each lunar module commander chose to opt for manual landing.[37] The last lunar module commander, Eugene Cernan, recalled his thoughts on switching to "manual" control: "There's no way I'm going to go all the way to the Moon … and let a computer land me."[38]

POSTSCRIPT

There is a final tale in the control room history of Apollo that we should layer with the fashioned surface of the Playtex spacesuit.

The Apollo Guidance Computer (ACG) was the first computing device to use hardware similar to modern computers—silicon wafers containing multiple transistors, saving weight and elec-tricity while increasing computing power. Its software, however, was stored and loaded very dif-ferently from today's applications. As the modern user can attest, the back-and-forth of loading and executing instructions in a computer can be more than unpredictable. The Space Shuttle, designed only a few years after Apollo, carries a redundant suite of five identical IBM comput-ers.[39] The Apollo Command Module and lunar lander, by contrast, each held only a single such device. When the Command Module and Lunar Excursion Module were docked together for the long trip to the moon and back, the two computers could substitute for each other—as they did during the crisis of Apollo 13. But for the most nerve-racking episode of the lunar journey, the separation between the two craft, and the spidery LEM's descent to and ascent from the moon's surface, the mission's success relied on a single computer only.

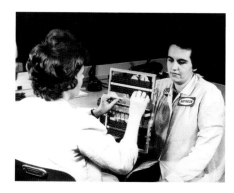

18.14

Hand-weaving of rope memory wiring for Apollo Guidance Computer, Raytheon, Waltham, Massachusetts.

To help ensure the reliability of the single device's software, MIT engineers charged with its manufacture turned away from systematic solutions. As a result, guidance computer manufacturing would employ some of the only other women, outside of Playtex, in the entire Apollo effort.

To allow the guidance computer's programs to reload instantly in the event of a software "crash," the engineers eschewed the ponderous tapes and disks used to load ground-based computer software. Instead, once a tape of the ACG's software was finalized for each mission (software was constantly updated throughout Apollo), the ones and zeroes of binary instructions were hand-crafted in fabric-like "ropes" of thin wire and magnetic washers. A current passed through the long rope would register each literal "bit" of ferrous metal along its length, and load the spacecraft's guidance software without the possibility of error. To make each rope precisely, however, was an enormous technical challenge. As an executive at Raytheon, responsible for the ACG's manufacture, proclaimed: "We have to build, essentially, a weaving machine."

According to an MIT manager on the project, Raytheon "tried building the components with supervisors, industrial engineers … [but] I mean, everything they made was scrap."[40] So, drawing from the textile mills and watch-makers around Raytheon's Waltham, Massachusetts factory, the firm hired a staff of women (termed "LOLs," or little old ladies, by the engineers) to perform the delicate work. Raytheon managers were so afraid the women would lose the knack of the delicate assembly work that they would pay them to sit and wait for weeks on end while each mission's software was finalized. The time was spent knitting.[41]

Notes

1. See Reinhold Martin, *The Organizational Complex: Architecture, Media and Corporate Space* (Cambridge, MA: MIT Press, 2003), 191.

2. In the Second World War, designers Raymond Loewy, Henry Dreyfuss, and Walter Dorwin Teague proposed a fantastical "Situation Room," complete with "Variable-Speed Statistical Visualizer," to the Office of Strategic Services (OSS). However, the scheme did not survive the transfer of jurisdiction of the OSS to the Joint Chiefs of Staff in 1942, and the situation rooms of the Second World War were confined to more limited architectures. The Visual Presentation Branch of the OSS, however, would become known for its information graphics, which included charts used at the Nuremberg trials, as well as the logo for the postwar United Nations. See Barry Katz, "The Arts of War: 'Visual Presentation' and National Intelligence," *Design Issues* 12, no. 2 (Summer 1996): 3–21.

3. Caption to Air Force photo 171235, National Archives RG 342-B-11-082.

4. Paul E. Ceruzzi, *The History of Modern Computing* (Cambridge, MA: MIT Press, 2003), 124.

5. Charles McKinney, "Mission Control Center—Houston Display and Control System," *Supplement to IEEE Transactions on Aerospace*, June 1965, 121–125; also "MCC Development History," document of the UHCL/JSC archive, compiled by Ray Loree, August 1990.

6. McKinney, "Mission Control Center."

7. Architectural historian Beatriz Colomina considers "Glimpses of the USA" in her article "Information Obsession: the Eameses' multiscreen architecture" (*Journal of Architecture* 6 [Autumn 2001]), linking it to the origins of "array[ed], multiple images," in parallel with "situation rooms," and NASA's "Cape Canaveral … Mission Control."

8.   Loree, "MCC Development History."

9.   So termed by Mercury flight controllers; see R. A. Hoover, "Apollo Experience Report: Flight-Control Data Needs, Terminal Display Devices, and Ground System Configuration Requirements," NASA Center for AeroSpace Information (CASI) NASA-TN-D-7685; JSC-S-396, 19740501, May 1, 1974, 7.

10.   Loree, "MCC Development History."

11.   *Architectural Design*, February 1967, 89.

12.   F. P. Brooks, Jr., G. A. Blaauw, and W. Buchholz, "Processing Data in Bits and Pieces," *IRE Transactions on Electronic Computers* EC-8 (1959): 118–124; reprinted as "Architectural Philosophy," in Werner Buchholz, ed., *Planning a Computer System: Project Stretch* (New York: McGraw-Hill, 1962), 5–17.

13.   Flight Operations Division, Manned Spacecraft Center, Langley Station, Hampton, Virginia, "Plan for Design, Implementation and Operation of Flight Control Complex for Projects Gemini and Apollo," March 29, 1962 (UHCL/JSC Manned Spaceflight Article, Box 7).

14.   International Business Machines Corporation, "Introduction to the Manned Spaceflight Control Center," January 10, 1965, UCHL/JSC Manned Spaceflight Archive (Mission Control Collection, Box 7).

15.   "Mission Control: FIDO, GUIDO, and RETRO," *Time* 94, no. 5 (August 1, 1969).

16.   Ceruzzi, *The History of Modern Computing*, 124.

17.   Lloyd Steven Sieden, *Buckminster Fuller's Universe: His Life and Work* (Cambridge, MA: Perseus Publishing, 1989), 53–59.

18.   Arthur Ferraro, NASA MSC news conference, MIT, Cambridge, MA, September 24, 1963 (University of Houston-Clear Lake JSC Archive, Chronographic File 193-64, 21); quoted in David Mindell, *Digital Apollo: Human and Machine in Spaceflight* (Cambridge, MA: MIT Press, 2008), 108.

19.   Walter McDougall, *The Heavens and the Earth: A Political History of the Space Age* (Baltimore: Johns Hopkins University Press, 1997), 449.

20.   National Aeronautics and Space Administration, *Apollo Program Summary Report*, NASA Document JSC-09423 (Houston: Lyndon B. Johnson Space Center, April 1975), 4–108.

21.   Jim Lovell and Jeffrey Kluger, *Lost Moon: The Perilous Voyage of Apollo 13* (Boston: Houghton Mifflin, 1994), 270.

22.   Henry S. F. Cooper, "A Reporter at Large: Life in a Space Station," *New Yorker*, August 30, 1976, 44.

23.   Ibid., 52.

24.   Reuters wire service, "Lethargy of Skylab 3 Crew Is Studied," *New York Times*, December 11, 1973, 14.

25.   Johnson Space Center Oral History Project, transcript of interview with Gerald P. Carr, conducted by Kevin M. Rusnack, Huntsville, Arkansas, October 25, 2000, 12–48.

26.   Mary Lou E. Balbaky with Michael B. McCaskey, "Strike in Space," HBS No. 9-481-008 (Boston, MA: Harvard Business School Publishing, 1981). In an interview with the Johnson Space Center Oral History Project, astronaut Edward Gibson contested the HBS account, arguing that the crew had mistakenly turned off all of the radios. Gerald Carr's last communication to Mission Control before the communication gap, however, reads much like a manifesto: "We had told [you] on the ground

before we left that we were not going to allow ourselves to be rushed; yet we got up here and let ourself just get driven into the ground … I think really up here my biggest concern is keeping the three of us alert and healthy." Mission transcript reported in Henry S. F. Cooper, *A House in Space* (New York: Holt, Rinehart and Winston, 1976), 130; see also Johnson Space Center Oral History project, interview with Edward G. Gibson, conducted by Carol Butler, Houston, Texas, December 1, 2000.

27. Cooper, "A Reporter at Large," 49.

28. Ibid., 37.

29. Ibid., 43.

30. Ibid., 63.

31. Cooper, *A House in Space*, 12.

32. United States, *Briefing by Skylab IV Astronauts: Hearing before the Committee on Science and Astronautics, House of Representatives, Ninety-Third Congress, Second Session, April 23, 1974* (Washington, D.C.: U.S. Government Printing Office, 1974), 42.

33. Ibid., 16.

34. Ibid., 15–39.

35. Associated Press, "Astronauts Report Orbital Flight Alters Attitudes about Life," *New York Times*, January 3, 1974, 74.

36. See Mindell, *Digital Apollo*, 252.

37. Ibid., 269.

38. NASA Johnson Space Center Oral History Project, Oral History transcript, Eugene A. Cernan interviewed by Rebecca Wright, Houston, Texas, December 11, 2007, 2.

39. Four of the five identical IBM 4Pi/AP-101 computers run the Shuttle guidance software. The devices will outvote each other if one reaches an erroneous conclusion, or crashes in the process. A fifth computer runs a completely different software package in order to ensure that a systematic, software-based error does not leave the orbiter stranded. During earlier missions a disconnected, sixth, spare computer was kept on board as well. See Dennis R. Jenkins, *Space Shuttle: The History of the National Space Transportation System, the First 100 Missions* (Cape Canaveral, FL: D. R. Jenkins, 2001), 405–408.

40. Ed Blondin, Apollo Guidance Computer Conference 3 group interview, November 30, 2001, Dibner Institute of Science and Technology, History of Recent Science and Technology Program, <http://resolver.caltech.edu/CaltechAUTHORS:BUChrst06> (accessed December 3, 2008).

41. Jack Poundstone, Ed Blondin, Apollo Guidance Computer Conference 3 group interview, November 30, 2001, Dibner Institute of Science and Technology, History of Recent Science and Technology Program, <http://resolver.caltech.edu/CaltechAUTHORS:BUChrst06> (accessed December 3, 2008).

Layer **19**   CITIES AND CYBORGS

"The techniques that are going to put a man on the Moon are going to be exactly the techniques that we are going to need to clean up our cities," declared Hubert H. Humphrey in remarks at the Smithsonian in 1968.[1] The vice president's remarks were no simple comparison of difficulty; rather, they were indicative of a deliberate effort to shift the techniques, staffing, and even equipment of "Space Age Management"[2] to urban, and architectural, problems. Here on earth, however, the fissures and fragmentation separating systems engineering from human tissue would become as visible, if not more so, in space above.

THE TOTAL COMMUNITY CONCEPT

The notion that expertise gained in NASA should be extended to the operation of government and industry as a whole was an article of faith among NASA's founders. "We are going to spend 30–35 billion dollars pushing the most advanced science and technology," NASA administrator James Webb wrote on taking office in 1961, and are "endeavoring in every way possible to feed back what we learn into the total national economy."[3]

From the outset of the space race, Webb argued the importance of space-borne ideas for the social and physical landscape of the nation, emphasizing "the best ways of utilizing the tremendous developments of science and technology in what might be called a total-community, workable-plan kind of concept."[4] "It is incumbent on us, while achieving our specific mission objectives," Webb reminded colleagues in 1962, "to make available to citizens generally the specific practical benefits which can flow from a research and development program of this magnitude."[5]

The spread of such optimism is characterized by a 1968 *Science* editorial: "In terms of numbers of dollars or of men, NASA has not been our largest national undertaking, but in terms of complexity, rate of growth, and technological sophistication it has been unique. It may turn out that the most valuable spin-off of all will be human rather than technological: better knowledge of how to plan, coordinate, and monitor the multitudinous and varied activities of the organizations required to accomplish great social undertakings."[6]

Early in the decade, such logic had already been rendered explicit: in June 1962, thirty-five NASA officials and military-industrial executives were invited by Wayne Thompson (city administrator of Oakland, California, and president of the International City Managers' Association) to discuss areas of shared interest; in particular: "Can a national program of space exploration be applicable to the daily tasks of the men and women who live and work in our central cities? [And] how may [such] new knowledge … be used to seek answers to the critical issues facing expanding urban populations?"[7]

Out of this 1962 meeting came a March 1963 conference: "Space, Science, and Urban Life." It was supported by NASA and the Ford Foundation, in cooperation with the University of California and the City of Oakland. "It is imperative," declared Oakland manager Thompson in the conference's introduction, "that we depart from time-worn traditions and concepts and adopt space-age techniques to cope with the problems of our space-age cities."[8]

At the 1963 conference, the specific nature of such techniques was the subject of presentations by figures ranging from federal luminaries James Webb and Jerome Wiesner to Martin Meyerson, director of the Harvard-MIT Joint Center for Urban Studies, and Burnham Kelly, dean of the School of Architecture at Cornell. While discussions in the meeting ranged from communication to automation and biotechnology, the assembled group gravitated toward "systems analysis and management techniques" as holding particular promise for urban problems.[9]

This conclusion—that urban problems could be best tackled through systems engineering methods, and particularly their use of information—contained folded within itself another, more fundamental assertion: that cities themselves could be understood, like complex weapons, as flows of information and feedback. Mirroring Kline and Clynes's reframing of the human body as readily susceptible to cybernetic analysis and enhancement, the city became understood as a cybernetic system which could be simulated, retuned, and amplified to better serve "space-age" reality.

CITIES AND CYBERNETICS

"A city is primarily a communication center, serving the same purpose as a nerve center in the body. It is a place where railroads, telephone and telegraph centers come together, where ideas, information and goods can be exchanged." So wrote Norbert Wiener in the December 18, 1950 *Life* magazine.[10] In collaboration with several MIT colleagues, Wiener's main contribution to the issue was a pictorial essay advocating better highways and ring roads, as well as dispersal of resources, to help guard U.S. cities against atomic attack. This pictorial essay was a small part of a very public debate in the late 1940s and 1950s advocating "dispersal" of American cities—defensive suburbanization—as a bulwark against atomic attack. While it was of secondary import to economic and social forces in the physical development of America's postwar suburbs,[11] the dispersal debate had a lasting impact on the culture and vocabulary of postwar planning. First, it introduced the notion of a collaborative cadre of experts shaping urban form—academic figures such as Wiener, but also government and industry elites. Secondly, it (re)introduced the notion of the city as a scientific subject, abandoning the exclusively biological models of urbanity common since the nineteenth century in favor of ideas of information flow and feedback—which Wiener's new science of cybernetics had shown to describe the behavior of both natural and manmade systems.[12]

By the early 1960s, texts such as *A Communications Theory of Urban Growth* and *The Cybernetic Approach to Urban Analysis* proclaimed the city as a literal "cybernetic system … an information handling machine."[13] Thomas O. Paine, head of General Electric's Center for Advanced Studies in Santa Barbara, chaired a panel at the Institute of Electrical and Electronics Engineers International Conference in 1966 on "The City of Tomorrow," contributing his own paper on "The City as Information Network."[14] (Paine would replace James Webb as NASA administrator in 1968.)

New institutions such as the Urban Systems Laboratory at MIT sought to extend the reach and specificity of the cybernetic urban model. As well as several recruits from RAND,[15] the center's faculty included Whirlwind pioneer Jay Forrester, soon to publish his own work on cybernetic urban simulation, *Urban Dynamics*.[16] While Forrester allowed that the text made little or "no reference to [previous] urban literature,"[17] the work deployed cybernetic simulations of complex urban processes to forcefully advocate new, "scientific" approaches to planning. While confining its scope to a macroscopic urban scale, the book was especially influential among architects interested in a "rational," systems-based approach to design in cities.[18]

THE VISUAL SUBSYSTEM BY MEANS OF COMPUTER-GENERATED IMAGES

One of the most literal translations of Apollo's techniques to urban environments serves also to illustrate the importance of such informational principles. Starting in 1966, the Urban Laboratory Project of the UCLA School of Architecture and Urban Planning, headed by Peter Kamnitzer, entered into collaboration with General Electric. Their work focused on adapting for earthbound use the landing simulation system developed by GE for lunar simulation, in particular to aid the simulation of the new kinds of urban environments created by cybernetically inflected urban redevelopment—highways, underpasses, and superblocks (see Layer 12). In the first demonstration of the collaboration, taking place in the lunar landing simulator at Houston's Manned Spacecraft Center in December 1967, "a test subject 'drove' along simulated city streets. By moving a joy stick [sic] … he directed his path through the urban environment, maneuvered into and through a tunnel, across an overpass, through an underpass, and beneath a building supported on open columns."[19]

While capable of showing only 240 polygon edges,[20] the system was modified by GE and Kamnitzer's UCLA team to show a thoroughly contemporary cityscape, including elevated highways and skyscrapers. Further funding for the project came not from city governments but from the US Department of Transportation, which termed the project's domain "The Visual Subsystem" of the highway-navigated landscape.

CITIES AND SPACECRAFT, 1966–1974

"Although comparing a city to a spaceship may seem absurd," ventured a 1967 joint NASA/HUD report, "both are inventions produced by men. The speed with which spaceships have been built makes many men wonder why progress in urban affairs has not been accelerated, too."[21]

The document, entitled *Science and the City*, was the result of a three-week summer session on "Science and Urban Development" at Woods Hole, Massachusetts, organized jointly by NASA and the newly formed Department of Housing and Urban Development (HUD) in the summer of 1966. The Woods Hole conference, as well as the action-oriented language of its resulting report, marked an essential shift. Instead of simply imagining a translation of the military-industrial systems approach, grounded in cybernetics, to urban policy, the 1967 report advocated the implementation of computer-driven "urban control systems." Unlike the 1963 meeting, where top NASA officials met with politicians and local officials, the 1966 conference included equal numbers of government, industrial, and federal representatives, and came at a time when ugly urban riots, public failures of previous "renewal" schemes, and congressional pressure led President Johnson to propose (in a message accompanying the Demonstration Cities and Metropolitan Development Act of 1966) "that we focus all the techniques and talents within our society on the crisis of the American city."[22]

A year earlier, Senator Gaylord Nelson had introduced a bill (S.2662) proposing that $125 million be provided to state and local governments "to design computer programs that would test various solutions for urban woes with the same logic used in designing moon rockets."[23] By 1968, RAND, the same Santa Monica think tank that had established the basis for Cold War simulation (and worked closely with Bernard Schriever and others in laying the conceptual foundations of the systems management), opened an office in Manhattan under exclusive contract to the administration of New York mayor John V. Lindsay.

Especially as late-1960s defense budgets sacrificed aerospace spending for the escalating costs of the war in Vietnam, the movement of NASA employees and contractors from the frontiers of space to the "urban frontier"[24] became a choreographed exodus. During the summer of 1971, the Department of Labor and HUD funded intensive summer schools at UC Berkeley and MIT to retrain aerospace employees in techniques of urban management. Organized by the National League of Cities and the US Conference of Mayors, the programs focused on "how to apply systems thinking in a new milieu."[25] Placement programs matched the programs' graduates to mayor's offices, city planning departments, and budget offices around the country; within a year, 80 percent of graduates were working in urban administration.[26]

USA

In June of 1968, George Washington University, RAND spinoff Systems Development Corp., and North American Rockwell (builder of the Saturn V) organized a "Conference on the Urban Challenge." Speaking to the group, Bernard Schriever laid out his own vision of systems thinking and urban expertise; retired from the Air Force, he announced that during his last years of service, "I began to realize that many of the problems we faced in the Air Force in the research, development, and production of our weapons systems were similar to the problems facing our cities. I also became convinced," Schriever continued, "that the solutions we devised in the Air Force to our technical and management problems were applicable to the problems of our cities."[27]

With Simon Ramo and Dean Wooldridge, Bernard Schriever had invented a new kind of organization in the 1950s—a complex military-industrial-academic collaboration, brought together by the "systems manager." Schriever, as commander of the Western Development Division, was the Ur-systems manager himself (see Layer 4). Having helped found the military-industrial complex, by 1967 he was assembling a new kind of collaboration, "a consortium of eight to ten leading companies … aimed at being able to do a city-system study and then to execute a program based on that study … a city rehabilitation."[28]

Schriever's partners in the (for-profit) "city-rehabilitation" venture included Lockheed, Northrop, Raytheon, and Control Data Corp. (which competed with IBM in the sale of mainframe computers). The consortium was titled Urban Systems Associates, or USA. The venture's origin can be traced to the General's enlistment, closely following his retirement from the Air Force in 1966, as a consultant to the Department of Housing and Urban Development, where he served as an advisor to HUD's founding secretary, Robert Weaver.

OPERATION BREAKTHROUGH

As with the urban enlistment of Bernard Schriever, some of the most literal translations of expertise and procedure from a military-industrial to a civic context came at HUD, the cabinet-level urban department founded by Lyndon Johnson in 1966. Even after the loss of the Great Society's sponsor in the White House, the transfer of military-industrial insight to HUD continued. "Creating a safe, happy city is a greater challenge than a trip to the moon," the Woods Hole report had proposed, "because urban housing is more complex than a rocket. … Its ever-changing problems, nevertheless, can be attacked in the same logical way we have gone about exploring the universe."[29]

19.1

"A model of a hypothetical space craft equipped with
the components of a nuclear rocket propulsion,"
a design by Harold Finger from the Lewis Research
Center, 1961. Finger's work on nuclear-propelled
interplanetary spacecraft would form the basis of
the *Discovery* spacecraft in Stanley Kubrick's *2001*.

In January 1969, Harold Finger, NASA associate administrator for organization and management, was called to a late-night meeting at HUD. George Romney, Richard Nixon's HUD secretary, wished to invite Finger to join his administrative team.[30] Leaving NASA, Finger became HUD's first assistant secretary for urban technology and research in March.

Harold Finger had no previous experience in urban or social policy ("I really don't know how they [found] me," he later confessed).[31] Before joining NASA's administration in 1967, Finger had for a decade served as head of nuclear propulsion research at NASA, as well as its predecessor, NACA. While initial work had focused on nuclear propulsion for airplanes—including the astonishing installation of a working nuclear reactor in a Convair B-36 "peacemaker" bomber[32]—Finger's later research focused on nuclear-powered interplanetary spacecraft; these prototypes would be used by Stanley Kubrick to shape of the *Discovery* spacecraft in *2001*.[33]

Discussions around Finger's recruitment by HUD Secretary George Romney in late 1968 directly referenced "turmoil in the cities—riots, real problems,"[34] and the continued need for a "systems approach."

The largest project overseen by Finger was the multimillion-dollar "Operation Breakthrough." Instead of "piecemeal" solutions, Operation Breakthrough attempted to address the country's urban crisis through "new total systems for constructing and marketing" urban homes and communities.[35] Instead of segregating populations on the basis of race and income, Breakthrough sought to catalyze prefabricated, mixed-income housing construction that would both allow the U.S. housing industry to "break through" to innovations in standardization, systemization, and mass production, and further allow the country to "break through" its own sense of urban malaise.[36]

Despite his lack of specific qualification, Finger's expertise was embraced both by aerospace firms—who sought to enter the construction and transportation sectors—and by the building industry itself, where a fascination with the possibilities of "systems building" was one of the main legacies of the space program in American architectural culture.

Unlike their foreign colleagues—such as the British group Archigram—the American architectural press had elevated none of the physical artifacts of the space race as worthy of architectural imitation. Instead, the focus was on systems management, and its interface with prefabricated, physical architectural "systems.[37] Operation Breakthrough was especially representative; its goals were "the application of technology," and in particular "prefabrication, heavier mechanization … modular standards … integration," and the direct involvement of "aerospace corporations" and conglomerates "in revolutionizing the landscape of American building."[38]

To advance Operation Breakthrough, Finger recruited a range of personnel from military-industrial backgrounds, and actively consulted his former superiors such as James Webb and Thomas O. Paine.[39] Instead of being managed "helter-skelter,"[40] Operation Breakthrough was directed from an aerospace-style "Control Room" in Marcel Breuer's new HUD headquarters. There, the systems management of Bernard Schriever was directly applied to the creation of model American neighborhoods. Following an aerospace model of procurement, companies such as Alcoa, Bechtel, and Martin-Marietta proposed hundreds of prototype housing systems. Of these, 22 were funded for full development, including proposals by Lockheed, General Electric, and Simon Ramo's own company, TRW (the T had been added to Ramo-Wooldridge in a 1958 merger

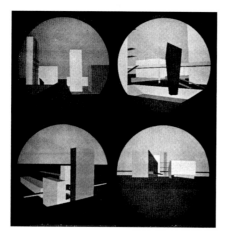

19.2

NASA's digital lunar landing simulator, built by GE, retrofitted to navigate a "cityscape."

19.3

Shared HUD/DOD computing facilities, deployed for site planning and system analysis during Operation Breakthrough.

with Thompson Products). In the first years of the program, these systems were used to build 2,800 units on nine test sites from Jersey City to Kalamazoo, incorporating innovations not only in prefabricated constructions, but also in site systems such as waste management and pneumatic trash collection. Particular emphasis was given to "blighted" areas.[41] Once site planning and fabrication were under way, shared computing facilities with the Department of Defense were used to develop and modify modular housing designs according to structural, tectonic, hydrological, and even social variables.[42]

THE *DE NOVO* CITY

Whereas Operation Breakthrough sought in principle to integrate its digitally inflected fabric into existing urban settings, a final set of proposals favored the creation of urban form from scratch.

"The *de novo* city is a very exciting concept," rhapsodized Litton Industries' Roy Ash in 1966. Citing "too many problems … all of them interrelated," in existing cities, Litton Industries called for "integrating the urban problem … raising it to a higher level," by "starting from scratch."

Tex Thornton's successor as Litton's president, Ash raised the prospect of Litton-built cities in a 1966 *Fortune* magazine interview, presenting it as the most exciting prospect for extending the firm's "systems management capability." "By 1970," he explained, "we think that we'll be able to build a city that is adequate for 25,000 and has prospects of growing to 500,000. By 1990 we will be designing cities for a million people with the prospect of growing to five million."

"Don't worry," Ash demurred, "the systems management work involved in getting a one-million-population city in business overnight is probably more sophisticated … than the work required to get to the moon." But, concluding the discussion, he reaffirms that "this whole business of military command-and-control systems has considerable potential."[43]

SLIPPERY WATER

Advocating *de novo* cities, Ash cited "too many" interrelated problems in existing contexts. Such a forest of interrelated difficulties would come to plague the actual outcome of system engineering in the city.

Despite trumpeted technological successes—such as the introduction of "slippery water" into the city's fire hoses[44]—the New York City Rand Institute, one of the most public collaborations between military-industrial management and urban government, met increasing resistance from both the government and the people of its urban host. At its peak, the Institute had commanded four floors of midtown office space, a staff of 60 consultants, and $75 million in city contracts. Starting in the early 1970s, however, it became subject to increasing criticism, both from city financial officials (such as Comptroller Abraham Beame, Lindsay's eventual successor) and from the many communities affected by the policies Rand shaped with its (often confidential) recommendations. There is a "growing unwillingness" on the part of communities, allowed Rand director Peter Szanton in 1970, to "accept decisions which affect their lives but which they have no part in making."[45] With the city council dismissing Rand reports as "curiously irrelevant"[46] to the practice of city government, contracts were reduced in scope; the Institute finally closed in 1974. (Also closing in 1974 was MIT's Urban Systems Laboratory, again for lack of funding.)[47]

The consortium founded by Bernard Schriever, Urban System Associates, encountered equal difficulty. While extended negotiations took place over its participation in a San Francisco urban redevelopment proposal and a regional water plan in Arizona, no consulting contracts were ever signed by the consortium, and Schriever—uncharacteristically—"admitted defeat," dissolving the organization in 1969. Reflecting later, he observed: "Though the need for a systematic approach to urban development exists, unfortunately its application today is extremely difficult. The key problem is that social and psychological phenomena cannot be rigorously posed in mathematical terms ... due [in part] to the lack of a meaningful data base."[48]

Yet much work continued, with the acquisition and interpretation of urban data its particular focus. Los Angeles, praised in 1964 by *The Cybernetic Approach to Urban Analysis* both for its "early and pioneering efforts to use automation in the form of computers as a tool of city planning" and for its "application of cybernetics to urban problem-solving,"[49] continued to fund a digital Community Analysis Bureau through the 1970s. And "urban systems divisions" continued in think tanks such as RAND and its spinoff Systems Development Corporation.[50] The legacy of this military-industrial involvement in today's urban information tools is palpable, although infrequently acknowledged.[51]

SURVEYING BREAKTHROUGH

Operation Breakthrough—the largest and most detailed project to implement systems management in the creation of urban fabric—met its own complex set of challenges.

The first of these were not technological but social and political. While several hundred cities initially volunteered to be sites for the experimental program, at least one finalist site was abandoned when community objections to the mixed-income component of Breakthrough's proposals became too vociferous.[52] Beyond local politics, the shift in construction proposed by Breakthrough—from on-site work divided by craft, to assembly-line efforts by mostly nonunion labor—met fierce resistance from labor unions; the resulting political volatility led to some areas of the country being considered practically untenable for the proposed construction.[53]

Further problems were economical and technological in nature. While a significant impetus behind Breakthrough had been perceived cost inefficiency in existing, site-based construction, many of the manufactured housing systems produced for Breakthrough communities failed to accomplish any savings over conventional methods. Despite HUD Secretary George Romney's prediction that, within five years of Breakthrough, "two-thirds of all the housing built in this country will be basically factory-made,"[54] by 1976 only five of the 22 systems developed for Breakthrough were still being produced.

A final set of technological difficulties involved the advanced infrastructural systems installed on Breakthrough sites, in particular the sophisticated systems deployed to accommodate refuse. At the Jersey City Breakthrough site, a 1978 HUD study found that persistent faults in a pneumatic trash system resulted in a 46 percent failure rate, with predictably malodorous results.[55] The Nixon Administration shut down Operation Breakthrough a year later. (Harold Finger had departed HUD in 1972, becoming a spokesperson for General Electric's nuclear power business.)

Assessing the program in 1976, a General Accounting Office report concluded that Operation Breakthrough "did not accomplish its objectives."[56] Of the industry experts surveyed by the GAO, less than a third thought that Breakthrough had even made a "minor contribution."[57] (Decades later, however, it should be noted that the extension of aerospace techniques to building forecast by Breakthrough has become standard practice, at least in limited architectural contexts.)[58]

"THE SHATTERED IMAGE OF LITTON INDUSTRIES"

Litton Industries' plans to enter the city-building business had the most public failure, falling prey to a larger scandal over the distance between perception and reality in the company's systems management culture.

The December 1, 1969 issue of *Forbes* magazine featured a cover of broken glass, and the headline "The Shattered Image of Litton Industries." "Much of the glamour" of Litton, *Forbes* reported, "is gone. And," it added, "for good reason."[59] After almost 15 years of a continual rise in its earnings, Litton stunned Wall Street in January of 1968 with the revelation that earnings would be "substantially lower" than expected. The only public explanation for the sudden change in fortune was "earlier deficiencies of management personnel."[60]

The reverberations extended beyond Litton's own future, and were taken by Wall Street to signify that the conglomerate strategy pioneered by Litton did not actually lead to the efficiencies of scale, interrelationships, and "synergy" with which elaborate acquisitions had been originally promoted. "Litton was the sacred cow of these ... companies," one broker commented, "but yet they stumbled. Who's to say the others aren't equally vulnerable?" "What is clearer almost daily," *Forbes* lamented, "is the considerable distance between concept and reality at Litton."[61]

Despite Litton's public failings, however, the particular expertise of its management team continued to play a prominent role in public affairs. Roy Ash, author of Litton's city-building scheme, moved on to a central role in the Nixon administration. He was first appointed as the head of a Task Force on Government Reorganization that proposed a massive (and ultimately unsuccessful) "synergistic" rearrangement of the Executive Branch.[62] In the second Nixon term, Ash became director of the White House Office of Management and Budget; his "new, expanded role" was a sweeping mandate from Nixon to survey and evaluate "all Government programs now in existence," with a special focus on welfare and urban revitalization efforts.[63]

DEATH AND LIFE

This legacy of failure on the one hand, and continued influence on the other, makes the systems movement in architecture and urban planning particularly worthy of attention. The superficial parallels between systems approaches in architecture and planning, and the predilection of architects for hard, systematized spacesuits (see Layer 17), are themselves arresting. Beyond this surface lies an even more essential insight.

When, in the last chapter of her 1961 opus *The Death and Life of Great American Cities*, "The Kind of Problem a City Is," urban activist Jane Jacobs sought to articulate a metaphor for urban planning distinct from the "collection of file drawers"[64] she abhorred, she turned to recent work at the Rockefeller Institute, which had provided her with funding and an office to assemble her manuscript. There, her neighbor, Dr. Warren Weaver, was composing an essay for the foundation's 1958 annual report that identified new thinking in the natural sciences, which tried to understand what Weaver termed "organized complexity." (The frontispiece to Weaver's essay was a model of the newly discovered DNA structure.)[65] Heavily quoting from Weaver's essay, Jacobs makes a case for the special affinity between urban landscapes and complex biological systems. Invoking the example of a single urban park, she submits that any attempt to isolate the variables, however many, leading to the success or failure of an urban enterprise is inherently dubious; those variables are "too numerous and interconnected."

Contrast such an analysis with NASA and HUD's 1967's *Science and the City*: "The city, too, consists of systems and sub-systems. ... We have experts on the operation of social as well as physical *black boxes*. Their talents and skills, however, have usually been brought together on urban problems helter-skelter, rather than *by a systems approach*."[66]

By contrast, Weaver's 1958 essay on science and complexity was particularly circumspect when it came to envisioning "black boxes" in nature. The essay highlighted the problem of "dissectability": "Between the living world and the physical world," Weaver cautions, "there is a critical distinction as regards dissectability. A watch spring can be taken out of a watch and its properties usefully studied apart from its normal setting. But if a heart be taken out of a live animal, then there is a great limitation on the range of useful studies which can be made."[67]

As to the contrast between living and electromechanical systems studied *in situ*, Weaver offers a further caution that is itself, in retrospect, wildly optimistic: "The significant problems of living organisms are seldom those in which one can rigidly maintain ... variables. Living things are more likely to present situations in which a half-dozen, or even several dozen quantities are all varying simultaneously and in subtly interconnected ways. And often they present situations in which some of the essentially important quantities are either non-quantitative, or at any rate have eluded identification or measurement." While the insights of cybernetics on feedback and homeostasis help us understand the nature of nature, their actual function and complex relationships remain—as Jane Jacobs remarked of the dynamics surrounding a single urban space— "as slippery as an eel."[68]

While presciently dismissive of attempts to "reduce biology ... to physics,"[69] Weaver's mention of a daunting "dozen" of variables in a given natural system would prove in turn to be a significant oversimplification. The revolution in biological science foretold and funded by Weaver has only increased our appreciation for the proliferating complexity of nature.

19.4

A rendering from the 1975 NASA Ames study *Space Settlements*, showing modular building components deployed for residential use within the artificial gravity of a toroidal space station located at Lagrange point 5 between the moon and earth.

To take one example, Weaver himself refers to the newly discovered DNA molecule as the "blueprint of life"—implying, as many mistakenly believe, a systematic connection between base pairs and biological form directly analogous to the contractual documents governing architectural construction.[70] In reality (to borrow Weaver's own words), there is much that "eludes identification or measurement." Indeed, it has recently been asserted that "although mutational change [that is, change in genetic instructions] is needed for phenotypic change [change in the visible architecture of life], the two are simply not related."[71] Or, in other words, "Genes may have an effect on the phenotype, but this effect strongly depends on other genes."[72] Our own body's "blueprints," it transpires, do not perform directly, or even transitively, but rather through a complex web of shifting relationships, disturbed even by our own observation of them.[73] Such a web recalls not the city as we would design it but, rather, the confusion of a real building site and, indeed, the city itself: a fog of conflicting instructions and impressions, whose very vitality we may threaten through "systematic" attempts to fix it—let alone seek its reproduction.

"PARTICIPANT EVOLUTION"

In this context, we should see Wayne Thompson's 1963 exhortation to "adopt space-age techniques to cope with the problems of our space-age cities" as cut from the same conceptual cloth as Kline and Clynes's "cyborg." In reality—and in spacesuits and cities both—complex nature subverted such a systematic frame.

Whether body or city, attempts to reduce organic, "organized complexity" to single inputs and outputs fall prey to the same ambition as the militaristic acronyms that permeate such explanations. They flatten multiple complexities to create a world "atomized and redesigned … spare and letter-sleek."[74] As we have seen, the particularly unsleek and nonatomic quality of the 21-layer A7L spacesuit repeatedly threatened its own survival in the acronymic universe of the military-industrial space race. Yet, as we have also observed, it was the very complex, amorphous qualities of the suit's surface that allowed its occupants to survive the literal, and perhaps also organizational, space of the space race.

SPACE SETTLEMENTS

While Operation Breakthrough had, at best, a "minor" impact on the earthbound housing industry, a final footnote to its space-age origins is provided by a post-Apollo strategic plan for US space stations, *Space Settlements: A Design Study.* Produced by NASA's Ames Research Center (as a parallel activity to the AX speculative suit designs), the 1975 study identifies systems developed for Operation Breakthrough as "especially suitable for building in space."[75]

Drawing on the same space station research exploited by Kubrick's *2001*, the habitat proposed by Ames would house 10,000 astronauts, optimized by age and gender and "settled by persons from Western industrialized nations."[76] In the renderings accompanying the proposal, the components of systems building are seen in perhaps their only native soil: a perfectly managed, balanced world entirely subject to "principles from government and industry." As in a Renaissance tableau, it is only in a celestial realm that man steps above his complex nature.

Notes

1. "HHH on the Space Program," *Aerospace Technology* 21, no. 24 (May 20, 1968): 19. Excerpts from a speech given on May 7, 1968 at the Smithsonian Institution, Washington, on the awarding of the Robert J. Collier Trophy.

2. See James E. Webb, *Space Age Management: The Large-Scale Approach*, McKinsey Foundation lecture series (New York: McGraw-Hill, 1969).

3. James E. Webb, NASA Administrator, to E. F. Buryan, July 18, 1961, James E. Webb Papers, Harry S. Truman Library, Independence, MO.

4. James E. Webb, NASA Administrator, memorandum for Mr. Stoller, April 23, 1962, Webb Papers.

5. James E. Webb, NASA Administrator, memorandum for Directors, NASA Field Centers, Western Operations Office and North Eastern Office, The Industrial Applications Program, September 19, 1962, Webb Papers.

6. Wolfe Dael, "The Administration of NASA," *Science* 163 (1968): 753.

7. National Aeronautics and Space Administration, *Conference on Space, Science, and Urban Life, Proceedings of a Conference Held at Oakland, California, March 28–30, 1963, Supported by the Ford Foundation and the National Aeronautics and Space Administration in Cooperation with the University of California and the City of Oakland* (Washington, D.C.: Office of Scientific and Technical Information, National Aeronautics and Space Administration, 1963), ix.

8. Ibid., 1.

9. A cogent analysis and context is provided by Jennifer S. Light, *From Warfare to Welfare: Defense Intellectuals and Urban Problems in Cold War America* (Baltimore: Johns Hopkins University Press, 2003), 111–112. See also R. Launius, "Managing the Unmanageable: Apollo, Space Age Management and American Social Problems," *Space Policy* 24 (2008): 158–165.

10. Norbert Wiener, Karl Deutsch, and Giorgio De Santillana, "The Planners Evaluate Their Plan," analysis of the "Wiener defense plan for cities," *Life* 29, no. 25 (December 18, 1950): 85.

11. See Light, *From Warfare to Welfare,* chapter 3: "Cybernetics and Urban Renewal"; also Peter Galison, "War against the Center," in Antoine Picon and Alessandra Ponte, eds., *Architecture and the Sciences: Exchanging Metaphors*, Princeton Papers on Architecture, 4 (New York: Princeton Architectural Press, 2003).

12. Light, *From Warfare to Welfare.*

13. Richard Meier, *A Communications Theory of Urban Growth* (Cambridge, MA: Joint Center for Urban Studies, 1962). At the time, Meier was a former nuclear chemist turned planner based at the Harvard-MIT Joint Center for Urban Studies. He later moved to UC Berkeley's College of Environmental Design, where, in his former office, now the author's, these words are being typed. Leland M. Swanson and Glenn O. Johnson, eds., *The Cybernetic Approach to Urban Analysis* (Los Angeles: University of Southern California, Graduate Program in City and Regional Planning, 1964), 10.

14. Institute of Electrical and Electronics Engineers, *IEEE International Convention Record* (New York: Institute of Electrical and Electronics Engineers, 1966), vol. 8, 147.

15. Such as Robert Levine and Thomas Schelling.

16. Jay W. Forrester, *Urban Dynamics* (Cambridge, MA: MIT Press, 1969).

17. Ibid., x.

18. *Architectural Design* (London), November 1971.

19. Slade Frank Hulbert and Peter Kamnitzer, *Highway Transportation System Simulation: The Visual Subsystem by Means of Computer-Generated Images. Prepared for the United States Department of Transportation, National Highway Safety Bureau* (Los Angeles: Reports Group, School of Engineering and Applied Science, University of California, 1969), I-4.

20. William J. Mitchell, *Computer-Aided Architectural Design* (New York: Petrocelli/Charter, 1977), 364. Here the GE system itself is misdated to 1967. A press release on the GE image simulator, including photographs of the system in operation, was released by NASA in 1964; an image from it appears as figure 12.10 in this book.

21. U.S. Department of Housing and Urban Development, *Science and the City* (Washington, D.C.: Government Printing Office, 1967).

22. Lyndon B. Johnson, Special Message accompanying the Demonstration Cities and Metropolitan Development Act of 1966, <http://www.presidency.ucsb.edu/ws/index.php?pid=27682> accessed December 12, 2009.

23. Cited in Light, *From Warfare to Welfare*, 119. Senator Nelson's full remarks appear in *Congressional Quarterly*, October 18, 1965. He would go on to be the sponsor of the legislation instituting Earth Day.

24. U.S. Department of Housing and Urban Development, *Science and the City*, 8.

25. Arthur Naftalin and Richard W. Gable, Aerospace Orientation Program, University of California, "Adapting Professional Manpower from Aerospace to Urban Government: Project Syllabus," Berkeley, CA, August 6–September 3, 1971.

26. William L. C. Wheaton, Warren W. Jones, and Warren H. Fox, *Adapting Professional Manpower from Aerospace to Urban Government: Final Report, Aerospace Orientation Program* (Springfield, VA: National Technical Information Service, 1972). See also Light, *From Warfare to Welfare*, 121.

27. General Bernard A. Schriever (ret.), "Rebuilding Our Cities for People," transcript of remarks given at the Conference on the Urban Challenge, June 19–21, Warrenton, VA, *Air Force and Space Digest* 51, no. 8 (August 1968): 64.

28. Ibid., 66.

29. U.S. Department of Housing and Urban Development, *Science and the City*.

30. Finger was invited to choose between this position and one equivalent to his administrative role at NASA. Harold Finger, interview with the author, January 11, 2010 (audio recording).

31. "Oral History Transcript, Harold B. Finger, Interviewed by Kevin M. Rusnak. Chevy Chase, Maryland—16 May 2002" courtesy NASA History Office, 61.

32. As noted in Layer 7, this 1954 installation placed 1,000-kilowatt nuclear reactor and massive, lead-lined crew compartment in a B-36 airframe. The compartment had lead and rubber walls several feet thick, and pilots gazed through foot-thick yellow protective glass. The light from this glass in turn gave a sickly cast to the standard gray interior of the plane, which was in turn painted lavender to compensate. See Dennis R. Jenkins, *Convair B-36 "Peacemaker,"* Warbird Tech Series, 24 (North Branch, MN: Specialty Press, 1999). Also Nicholas de Monchaux, interview with Harold Finger, January 11, 2010.

33. Harold Finger, interview with the author; also Frederick Ordway III, interview with the author, June 15, 2006 (audio recording). Ordway, former assistant to Wernher von Braun at NASA's Marshall Spaceflight Center, was hired as an assistant to Kubrick on the production.

34. Ibid., 62.

35. Promotional material, Operation Breakthrough, National Archives Record Group 207-HUD-MPF, Still Pictures and Photograph Collection, National Archives at College Park, College Park, MD.

36. For an extended discussion of this theme, see Colin Davies, *The Prefabricated Home* (London: Reaktion Books, 2005). Davies briefly discusses Operation Breakthrough on pages 82–83.

37. See for example *Architectural Design*, November 1971, special issue on "systems approach to building" edited by Building Systems Design, Inc. Ezra Ehrenkrantz, founder of the firm, had attended the 1963 Oakland "Space, Science, and the City" conference.

38. *The Impact of Social and Technical Change in Building: A Report to the National Bureau of Standards, the Institute for Applied Technology* (Washington, D.C.: Building Systems Development, 1967).

39. Harold Finger, NASA Oral History Interview, 62. Aiming to preserve good relations, Finger favored military and aerospace personnel in staffing over his former colleagues at NASA. Also Harold Finger, interview with the author, January 11, 2010.

40. Ibid.

41. U.S. Department of Housing and Urban Development, *Operation Breakthrough … Now* (Washington, D.C.: Government Printing Office, 1974).

42. See Llewelyn-Davies Associates, *Operation Breakthrough: An Evaluation Framework for Site Planning* (New York: Llewelyn-Davies Associates, 1970).

43. Daniel Seligman and T. A. Wise, interview with Litton Industries president Roy Ash and senior vice president Harry Gray, "How Litton Keeps It Up, the View from Inside," *Fortune* 74, no. 4 (September 1966): 152–153, 180–182.

44. "Councilmen Ask Changes in Hiring of Consultants," *New York Times*, October 29, 1970, 1.

45. "The Men Who Tell City How to Run the City," *New York Times*, July 8, 1970, 40.

46. "Garelik Calls Rand Study of City's Police a Failure," *New York Times*, October 7, 1970, 55.

47. Light, *From Warfare to Welfare*, 121.

48. Paul Dickson, *Think Tanks* (New York: Atheneum, 1971), 216.

49. Swanson and Johnson, *The Cybernetic Approach to Urban Analysis*, iv.

50. Light, *From Warfare to Welfare*, 121.

51. On the unacknowledged military-industrial history of geographic information systems, or GIS, see Neil Smith, "History and Philosophy of Geography: Real Wars, Theory Wars," *Progress in Human Geography* 16 (1992): 257.

52. Harold Finger, interview with author.

53. United States, *Operation Breakthrough: Lessons Learned about Demonstrating New Technology, Department of Housing and Urban Development, Department of Commerce: Report to the Congress* (Washington, D.C.: General Accounting Office, 1976), 25.

54. U.S. Department of Housing and Urban Development, *Operation Breakthrough … Now*, 2.

55. Jack Preston Overman, Terry G. Statt, and David A. Kolman, *Operation Breakthrough Site Waste Management Systems and Pneumatic Trash Collection*, Utilities Demonstration Series, 4 (Cincinnati, Ohio: Environmental Protection Agency, Office of Research and Development [Office of Energy, Minerals, and Industry], Municipal Environmental Research Laboratory, 1978).

56. U.S. Department of Housing and Urban Development, *Operation Breakthrough … Now*, ii.

57. Ibid., 23.

58. The CATIA aerospace design system, manufactured by French Aerospace giant Dassault, was first used by architect Frank Gehry in the late 1980s; its use, and its derivatives, have spread widely; see William J. Mitchell, "A Tale of Two Cities: Architecture and the Digital Revolution," *Science*, n.s. 285, no. 5429 (August 6, 1999): 839, 841.

59. "Litton's Shattered Image," *Forbes*, December 1, 1969, 26.

60. *Wall Street Journal*, January 29, 1968, 25.

61. "Litton's Shattered Image."

62. "Nixon to Seek Restructured Government," *Washington Post*, January 22, 1971, A1.

63. "Shake-Up for the Team," *New York Times*, December 3, 1972, E1.

64. Jane Jacobs, *The Death and Life of Great American Cities* (New York: Vintage, 1961), 428–447.

65. Warren Weaver, "A Quarter-Century in the Natural Sciences," in Rockefeller Foundation, *Annual Report* (New York: Rockefeller Foundation, 1958), 1–91. See especially section 1 of the essay, "Science and Complexity."

66. U.S. Department of Housing and Urban Development, *Science and the City*.

67. Weaver, "A Quarter-Century in the Natural Sciences," 9.

68. Jacobs, *The Death and Life of Great American Cities*, 433.

69. Weaver, "A Quarter-Century in the Natural Sciences," 37.

70. See, for example, *US News & World Report*, special report on Craig Venter, "The Blueprint of Life" (October 31, 2005).

71. Marc Kirschner and John Gerhart, "Evolvability," in *Proceedings of the National Academy of Sciences of the United States of America* 95 (July 1998): 8, 420–427.

72. J. de Visser, G. M. Argan, et al., "Perspective: Evolution and Detection of Genetic Robustness," *Evolution: International Journal of Organic Evolution* 57, no. 9 (September 2003): 1959.

73. For example, the use of X-ray crystallography, discussed at length by Weaver as a revelatory technique, led, through the necessity of solidifying DNA into crystals, to the mistaken belief that a single genetic strand produced a single protein form. Recent advances in both microscopy and simulation have shown this to be far from the case. See Joram Piatigorsky, *Gene Sharing and Evolution: The Diversity of Protein Functions* (Cambridge, MA: Harvard University Press, 2007).

74. Don DeLillo, *Underworld* (New York: Scribner, 1997), 623.

75. Richard D. Johnson and Charles Holbrow, eds., *Space Settlements: A Design Study*, NASA SP-413 (Washington, D.C.: Scientific and Technical Information Office, National Aeronautics and Space Administration, 1977), 49.

76. Ibid., 51.

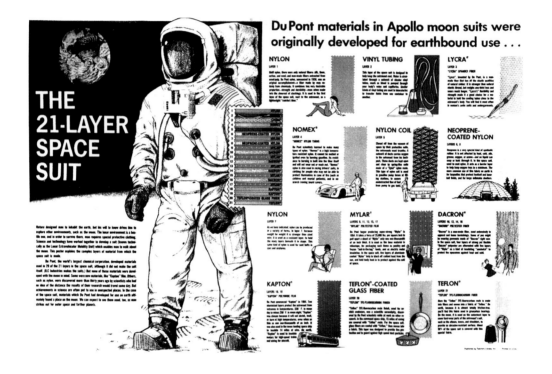

20.1

Promotional pamphlet showing use of
DuPont materials in 20 of the 21 layers
of the Apollo A7L.

"The 21 Layer Space Suit: DuPont materials in Apollo's moon suits were originally developed for earthbound use," proclaims a 1970 advertisement for the Delaware chemical giant. As the promotion illustrates, none of the materials used in the final suit was created *de novo*. Instead, they were appropriated for contexts—such as firefighting, in the case of Nomex, or lubrication, in the case of Teflon—whose performance criteria suited one of the manifold perils facing the astronaut's body. In the advertisement, DuPont uses the diversity of the pressure suit's materials to emphasize the seeming breadth of its own innovation: a universe of synthetics. Yet a more intriguing reality lies below.

As a 2002 self-published history acknowledged, DuPont's success in material innovation came "not only because its products were so diverse, but, more fundamentally, because their underlying chemistry was so similar."[1] As it sought to diversify from wartime production after the First World War (DuPont had produced gunpowder on the banks of Delaware's Brandywine since 1802), it was the discovery of a diversity of compounds sharing "chemical kinship"[2] that allowed the company to prosper in peace. Under the broad claim of "better living through chemistry," DuPont's diversity of products was carefully moderated within a consolidation of production processes, and narrowly confined research into defined chemical "families." For all the "New Look of Nylon" (so marketed in 1950), DuPont's material experiments shared the quality of Dior's own "New Look": a carefully rearranged, and repackaged, version of the old. Indeed, where *de novo* materials were attempted in the spacesuit's design—such as the Teflon-coated fiberglass cloth used for the suit's white covering—the results were not satisfactory, and a more quotidian composite had to be constructed.[3]

That novelty depends on precedent, and exists within a confined structure of innovation, should not be a startling proposition in a discussion of design. Yet especially in a context where discussions of history, or its absence, are often highly politicized, the remixed, reassembled layers of the A7L point toward a different kind of precedent: the logic of robustness and evolvability proposed by recent discussions in the natural sciences. Such a concept, of robustness as the context for natural innovation, also provides valuable insight into the consistent failure of systems engineering in natural systems—and into the peculiar fragility of the human body in space.

In 1958, Nathan S. Kline and Manfred Clynes proposed the cyborg to the Air Force as a call to a new kind of human development. "The challenge of space travel to mankind," they exhorted, "is not only to his technological prowess, it is also a spiritual challenge to take an active part in his own evolution."[4] This vision—to borrow Michael Sorkin's characterization of the Ames AX-5, the body "augmented and housed"—contradicts sharply the intimate realities of the soft, multilayered Apollo suit.

PANGLOSSIAN PARADIGMS AND EXAPTIVE EXTENSIONS

"Everything is made for the best purpose. Our noses were made to carry spectacles, so we have spectacles. Legs were clearly intended for breeches, and we wear them." So paleobiologist Stephen Jay Gould quotes Voltaire, and *Candide*'s Dr. Pangloss, in cautioning his colleagues against seeing perfection in nature.[5] In the same 1979 essay, "The Spandrels of San Marco and the Panglossian Paradigm: A Critique of the Adaptationist Programme," Gould deploys an architectural metaphor

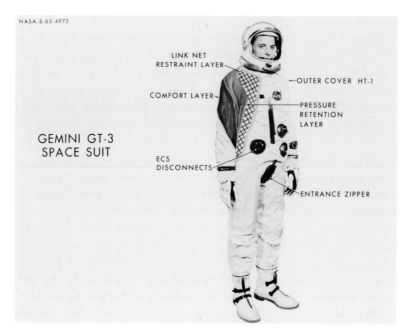

NASA-S-65-4973

GEMINI GT-3
SPACE SUIT

LINK NET
RESTRAINT LAYER

COMFORT LAYER

ECS
DISCONNECTS

OUTER COVER HT-1

PRESSURE
RETENTION
LAYER

ENTRANCE ZIPPER

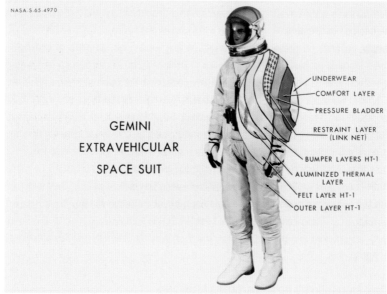

NASA-S-65-4970

GEMINI
EXTRAVEHICULAR
SPACE SUIT

UNDERWEAR
COMFORT LAYER
PRESSURE BLADDER
RESTRAINT LAYER
(LINK NET)
BUMPER LAYERS HT-1
ALUMINIZED THERMAL
LAYER
FELT LAYER HT-1
OUTER LAYER HT-1

20.2

Cross section, Gemini GT-3 suit.

20.3

Cross section, Gemini EVA configuration.

for the same idea. To assume perfection or optimization in a natural system, he argues, is just as misguided as imagining that the triangular spandrels of San Marco's dome (more properly pendentives) were precisely placed to accommodate triangular paintings.

We should then also take particular care against assuming that the 21 layers of the Apollo A7L are any kind of optimal number (especially given the conceit of this text's own structure). With the preceding profusion of layers in Gemini pressure suits (which went from three to twelve layers over the course of ten manned missions), and the elaboration of the 21-layer A7L into the 28-layer A7LB from Apollo 15 onward, we should see the layering of the A7L suit as not just a structure but a strategy and process.

Far from emphasizing optimization, recent work in evolutionary biology has stressed the role of "compartmentation, redundancy, robustness and flexibility" in the optimization of natural systems through evolution.[6] In the words of molecular biologist Sean B. Carroll: "one of the most important features that has facilitated the evolution of plant and animal complexity … is the modularity of their construction from reiterated, differentiated parts."[7]

Especially when stripped of its thermal micrometeoroid covering, the A7L spacesuit shows little of the fluidity and grace we have come to associate with "natural" design. In its tire-adapted bladder, girdle-adapted joints, and couture-adapted assembly, however, the A7L recalls especially one of the alternative modes of natural change advanced by Stephen Jay Gould, along with Elisabeth S. Vrba, to replace a purely "adaptationist" model of evolution—i.e., one that understands all natural form as necessarily perfected for its current task.[8] In a 1982 article, the two proposed the neologism "exaptation" to precisely describe evolutions in nature that are derived from traits originally developed for a quite different use. From eardrums (ex-reptilian jawbones) to wing feathers (duvets for freezing dinosaurs), "exaptations" far outnumber pure adaptations in our ecology. Even when a trait, like our own skeleton, seems instrumentally conceived for its current use, there is likely still a surprising shift in its history; for vertebrate bones, we must thank our Cambrian ancestors for laying up pockets of phosphates inside the body for metabolic purposes. Such deposits were adapted—or, properly, exapted—for skeletal use many millions of years later.

Subversive of our natural propensity to see reason in nature, the notion of exaptation is a particularly useful in understanding how complex natural systems like eyes, and wings, can arise serially from diverse antecedents. Much like the physical evolution of the Apollo spacesuit, "[t]he evolutionary history of any complex feature will probably include a sequential mixture of adaptations, primary exaptations, and secondary adaptations"[9] to reach its final form.

MODUS VIVENDI

In 1958, Clynes and Kline proposed a clean break with evolutionary history: "In the past, the altering of bodily functions to suit different environments was accomplished through evolution. From now on," they argued, "this can be achieved without alteration of heredity by suitable biochemical, physiological, and electronic modification of man's existing modus vivendi." Far from a *modus vivendi*, the 1958 cyborg proposal—as with much of the substance of the space race—followed a *modus operandi*. While systematic, it was a world quite distinct from the complexities of nature.

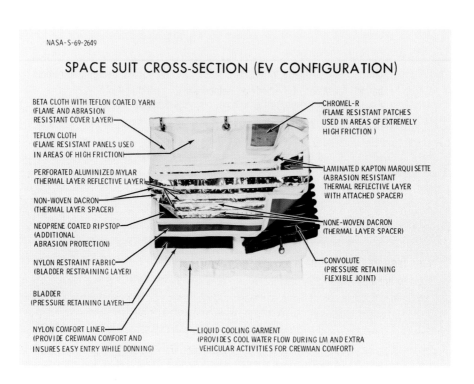

NASA-S-69-2649

SPACE SUIT CROSS-SECTION (EV CONFIGURATION)

BETA CLOTH WITH TEFLON COATED YARN
(FLAME AND ABRASION
RESISTANT COVER LAYER)

TEFLON CLOTH
(FLAME RESISTANT PANELS USED
IN AREAS OF HIGH FRICTION)

PERFORATED ALUMINIZED MYLAR
(THERMAL LAYER REFLECTIVE LAYER)

NON-WOVEN DACRON
(THERMAL LAYER SPACER)

NEOPRENE COATED RIPSTOP
(ADDITIONAL
ABRASION PROTECTION)

NYLON RESTRAINT FABRIC
(BLADDER RESTRAINING LAYER)

BLADDER
(PRESSURE RETAINING LAYER)

NYLON COMFORT LINER
(PROVIDE CREWMAN COMFORT AND
INSURES EASY ENTRY WHILE DONNING)

CHROMEL-R
(FLAME RESISTANT PATCHES
USED IN AREAS OF EXTREMELY
HIGH FRICTION)

LAMINATED KAPTON MARQUISETTE
(ABRASION RESISTANT
THERMAL REFLECTIVE LAYER
WITH ATTACHED SPACER)

NONE-WOVEN DACRON
(THERMAL LAYER SPACER)

CONVOLUTE
(PRESSURE RETAINING
FLEXIBLE JOINT)

LIQUID COOLING GARMENT
(PROVIDES COOL WATER FLOW DURING LM AND EXTRA
VEHICULAR ACTIVITIES FOR CREWMAN COMFORT)

20.4
Annotated cross section of A7L, 1969.

Reflecting in 1969 on the unsuitability of NASA-style management for use in urban administration (an opinion that differed strongly from that of his predecessor, Jim Webb, as well as his own earlier cybernetic speculations—see Layer 19), NASA administrator Thomas O. Paine described the difference between NASA and more chaotic structural orders. In contrast to the pure realm of technology, Paine characterized such settings as "Darwinian." Considering these "Darwinian Disciplines"—distinguished by "a yeasty ferment"—Paine concludes that "[m]obilizing modern science, technology and management to accomplish bold ventures in space is clearly far simpler than better organizing the extraordinary complex human interactions that comprise a modern metropolis."[10] And what is true of cities is true of bodies: if not yet quite reversing the historical tendency of architects and planners to compare cities to living bodies, contemporary biology has begun to abandon the vision of the body or metabolism as a mechanism.

Arguing for the existence of God in 1802, Scottish Reverend William Paley compared living systems to "watch-works," only more complex.[11] Even as it sought to blur distinctions between natural and manmade systems, the cybernetic model of the mammalian body deployed by Clynes and Kline in 1958 retained—in its equation of organs and engines—such a mechanistic world view. In today's study of natural systems, such a view is being supplanted by the idea of the body itself as a "mini-ecosystem, or economy," in which competing forces populate redundant systems to produce a far more complex broth of homeostasis than a clockwork, or cybernetic, model provides.[12]

ROBUSTNESS VERSUS REGULATION

Yet here an important question must be asked: were not the technological and organizational systems behind Apollo some of the most "enormously complex"[13] ever assembled by man? How can we effectively distinguish between the character of complexity clearly manageable through systems engineering innovations—the millions of bodies and billions of dollars flowing through Apollo's metaphorical veins—and the actual veins, arteries, and biological systems that constituted the astronauts at Apollo's center?

While the mathematical similarity and potential interweaving of the natural and manmade was a first principle of Kline and Clynes's cyborg, the qualities of successful biological and engineered systems turn out to have important differences as well.

With the foundation of the Air Force's Western Development Division in 1954, Bernard Schriever, Simon Ramo, and their colleagues set out to create management techniques that would allow the control of technological systems more complicated than any mankind had previously authored. The techniques they developed—systems engineering and management—would serve to both manage the complexity of such systems and, at an institutional level, create the superorganization of the military-industrial complex itself.

The legacy of systems engineering, as well as the program-planning budgeting system (PPBS) introduced by Robert McNamara in 1961 to help rein in development costs, was not so much their ability to survey complexity but, rather, to control it. The ability to control a complex architecture—through the organizational and operational "control rooms" that often blurred into one another—was the management system's goal, and to a great extent, its legacy.

To control complex systems and devices, organizations were created that mirrored precisely the technological structure of the systems they surveyed—and precisely not the more pedestrian logic of the rest of reality. Only by building a management system that explicitly mirrored the complex, cybernetic logic of the physical objects it assembled—Schriever's great innovation in the Atlas missile program—could the network of resulting organizations, military, industrial, and academic, succeed in the creation of complex missiles and the space program that developed from them. However, as has been proven dramatically, the logic of our own body (or ecology) is of a qualitatively, and even quantitatively, different order. As complex as the designed systems of the space race were, and as successful as the control mechanisms were that mirrored and shaped them, they were still enormously simple compared to the realities that sometimes thwarted them.

Before the introduction of systems management techniques in the Atlas missile program, small changes or perturbations in the missile's systems regularly produced catastrophic results. The quality of resisting perturbation—what biologists now call "robustness"—has come to characterize much recent discussion of life's inner mechanics.[14] Without qualities of robustness, complex systems, like a house of cards or Convair's first Atlas missiles, are highly prone to catastrophic failure. But the complexity—and resulting robustness—of life is infinitely more subtle than the most elaborate manmade system.

Consider, for a moment, the same "yeasty ferment" contrasted by Thomas O. Paine to NASA's system of "hierarchical controls." It has been shown that some 90 percent of a yeast's genes can be individually removed from the organism's chromosomes without a perceptible result. (The important caveat being that the lack of perceptible effect is true only under conditions of temperature and pressure that a yeast might consider "normal," if we can imagine such a reflective yeast.)[15] Similar resistance to mutational change has been observed in the genetic instructions of much more complex organisms like the fruit fly *Drosophila*, and in our own chromosomes as well. Such internal robustness, furthermore, has been shown to correlate closely to resistance to change in environmental conditions.[16]

We need only consider such examples as the errant spark of the Apollo 1 fire, or the effect of a single fused relay on the outcome of Apollo 13, to comprehend the relative lack of such "natural" robustness in the most complex manmade systems.[17] Yet to even imagine that the necessarily comprehensible technologies of the military-industrial moment could have approached a biological order of interdependence is to miss the nature of each system's origin. The blind search for a solution by an evolving biological system is likelier than not to turn up a robust solution;

20.5
Ethel Collins (left) and Sue Roberts (right)
retrofitting a pressure garment assembly.
Note paperwork in foreground.

that is, a solution represented by a large number of possible configurations (of a protein or RNA sequence, for example). In space-age engineering, the search for a solution from "first principles" was likely to come up with the opposite, a more singular result. One main interface between these opposites was skin, and spacesuit—the astronaut's own body.

Robustness and redundancy were central goals of the larger technologies of the Apollo program. But the nature of these qualities is very different at the biological and mechanical scales. Furthermore, the extreme limits of both the mission goals and limitations in Apollo, as well as the tendency of the numerically based systems (and later budget) management techniques to privilege quantitative over qualitative success, ensured that the solutions at the scale of the whole Apollo system would be antithetical to successful solutions at the scale of the astronaut's body.

The fragility of the human body in outer space, first explored in Layer 2, is here worthy of reconsideration. An essential qualifier to the robustness of a living system under internal (i.e., genetic) change is the stability of external conditions. And, while resistance to internal and external change has been correlated in real organisms, the extreme inversion of the human body's natural habitat in the extraplanetary environment creates out of its earthly resilience an astonishing fragility not nearly captured by the many threats listed by Clynes and Kline in 1958. Even if (to take the cyborg proposal seriously for a moment) technological mechanisms could be introduced to alter the operating temperature, pressure, and radiation level of the human organism, we can only imagine the countless, corollary adjustments necessary to the manifold, invisible mechanisms of human regulation, and the inevitable and unexpected side effects such pharmacological meddling would produce.

The A7L spacesuit was a solution to the problem not only of how to survive in space by ensuring livable pressure and temperature around the body, but the much more complex problem of how to make that survival robust to unanticipated changes, both inside and outside the spacesuit. Necessarily less complex than the evolved body it housed, it also had a manifestly different organizational quality than the engineered systems that surrounded it in turn.

Here the *fashioned* quality of the A7L, versus its many "first-principle" competitors, is essential. Derived from robust solutions for the body in space, and on earth, the modular and layered quality of the A7L, anathema to aerospace engineers, was particularly robust in accommodating the many inherently unpredictable challenges of suiting the body to space. This quality is characterized in the context of natural systems by systems biologist Andreas Wagner: "A system capable of fulfilling its primary function in many different configurations—explorable through mutation—has sufficient flexibility and degrees of freedom to adopt other features."[18]

A final, subtle, distinction returns us to our initial discussion of the 21-layer A7L's material quality. In contrast, for example, to the double bladder proposed by Hamilton Standard engineers to meet a systems management standard of reliability (see Layer 13), or deployed in the USSR's first EVA suit (see Layer 8), the multiple and interrelated layers of the A7L have a manifestly interdependent quality. To borrow again from a biological definition of robustness, "not redundancy of parts but distributed robustness is the major cause of robustness in living systems."[19] Considering the potential correlations between manmade and natural qualities of robustness, biologist Wagner himself suggests not only that "while it may be possible to design engineered systems simply and elegantly, robust engineered systems may be no simpler than biological systems." Particularly applicable to spacesuit design is Wagner's further observation that "the absence of completely rational, premeditated system design may favor the origin of distributed robustness."[20]

In this context, the innovation of the A7L was not the particular advances accomplished by using girdle-dipping techniques to produce convoluted joints, or even allowing underwear seam-stresses to deploy couture techniques to fuse disparate layers at 64 stitches to the inch. Rather, the underlying structure of the suit, with all its adaptive origins, softness, and layering, was the central innovation whose robust structure made subsequent, serial and additive innovations possible. The word *fashion*, we should remember, means not only a body of knowledge about clothing, but, as a verb, the transformation of one thing, usefully and often unexpectedly, into something else.

Notes

1. Adrian Kinnane, *From the Banks of the Brandywine to the Miracles of Science* (Wilmington, DE: E. I. du Pont de Nemours and Company, 2002), 111.

2. Ibid.

3. Kenneth S. Thomas and Harold J. McMann, *US Spacesuits* (New York: Springer / Praxis Books, 2006), 121.

4. Nathan S. Kline and Manfred Clynes, "Drugs, Space and Cybernetics: Evolution to Cyborgs," Symposium on Psychophysiological Aspects of Space Flight; Bernard E. Flaherty, *Psychophysiological Aspects of Space Flight* (New York: Columbia University Press, 1961).

5. S. J. Gould and R. C. Lewontin, "The Spandrels of San Marco and the Panglossian Paradigm: A Critique of the Adaptationist Programme," *Proceedings of the Royal Society of London*, Series B, Biological Sciences, 205, no. 1161 (1979): 21.

6. See Marc Kirschner and John Gerhart, "Evolvability," *Proceedings of the National Academy of Sciences* 95 (July 1998): 8420–8427.

7. Sean B. Carroll, "Chance and Necessity: The Evolution of Morphological Complexity and Diversity," *Nature* 409 (February 22, 2001): 1102.

8. Stephen Jay Gould and Elisabeth S. Vrba, "Exaptation—A Missing Term in the Science of Form," *Paleobiology* 8, no. 1 (Winter 1982): 4–15.

9. Ibid., 11–12.

10. Thomas O. Paine, "Space Age Management and City Administration," *Public Administration Review* 29, no. 6 (November-December 1969): 654–658.

11. William Paley, *Natural Theology, or Evidences of the Existence and Attributes of the Deity collected from the Appearances of Nature* (London, 1802), 2.

12. David Krakauer, "Cellular Struggle at Heart of Variety" (review of *The Plausibility of Life: Resolving Darwin's Dilemma* by Marc W. Kirschner and John C. Gerhart), *Times Higher Education Supplement*, November 27, 2006.

13. Thomas P. Hughes, *Rescuing Prometheus* (New York: Pantheon, 1998), 118.

14. See Andreas Wagner, *Robustness and Evolvability in Living Systems* (Princeton: Princeton University Press, 2005); Erica Jen, *Robust Design: Repertoire of Biological, Ecological, and Engineering Case Studies* (New York: Oxford University Press, 2005); and Gerhard Schlosser and Gunter P. Wagner, *Modularity in Development and Evolution* (Chicago: University of Chicago Press, 2004).

15. Victoria Smith, Karen N. Chou, Deval Lashkari, David Botstein, and Patrick O. Brown, "Functional Analysis of the Genes of Yeast Chromosome V by Genetic Footprinting," *Science*, n.s. 274, no. 5295 (December 20, 1996): 2069–2074.

16. Wagner, *Robustness and Evolvability in Living Systems*, 276.

17. To say nothing of the O-rings or insulating foam that doomed space shuttles *Challenger* and *Columbia*.

18. Wagner, *Robustness and Evolvability in Living Systems*, 218.

19. Ibid.,14.

20. Ibid., 315.

Layer 21 CONCLUSION

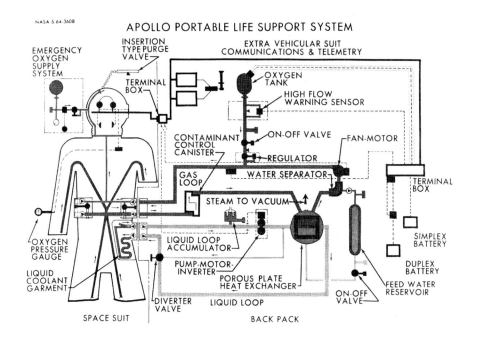

APOLLO PORTABLE LIFE SUPPORT SYSTEM

NASA-S-64-360B

EMERGENCY
OXYGEN
SUPPLY
SYSTEM

INSERTION
TYPE PURGE
VALVE

TERMINAL
BOX

EXTRA VEHICULAR SUIT
COMMUNICATIONS & TELEMETRY

OXYGEN
TANK

HIGH FLOW
WARNING SENSOR

ON-OFF VALVE

FAN-MOTOR

CONTAMINANT
CONTROL
CANISTER

REGULATOR

WATER SEPARATOR

GAS
LOOP

TERMINAL
BOX

STEAM TO VACUUM

OXYGEN
PRESSURE
GAUGE

LIQUID LOOP
ACCUMULATOR

SIMPLEX
BATTERY

LIQUID
COOLANT
GARMENT

PUMP-MOTOR-
INVERTER

POROUS PLATE
HEAT EXCHANGER

DUPLEX
BATTERY

FEED WATER
RESERVOIR

DIVERTER
VALVE

LIQUID LOOP

ON-OFF
VALVE

SPACE SUIT

BACK PACK

21.1

1964 NASA engineering diagram of the PLSS
and spacesuit.

Consider: first, a diagram of the Apollo life support system as conceived and illustrated by Hamilton Standard engineers in 1964; and secondly, an X-ray mobility study of the ILC A7L spacesuit performed by the Crew Systems Division of NASA in January of 1965. The fundamental differences between these two images return us to the question from which we began: Why—and perhaps now, how—was the Apollo spacesuit soft?

SYSTEMS AND CYBORGS

The first image is both uncharacteristic and singularly representative of Apollo's interleaving military-industrial machinery. It is representative of Apollo in that it presents the human body as a streamlined outline, graphically receding in favor of a system of inputs, outputs, and cybernetic relationships of control. It is uncharacteristic because the human body is shown at the most extreme apogee of Apollo's journey to the moon, in its lunar extravehicular activity (EVA). On the moon's surface, the body would be connected only to the portable life support system (or PLSS, shown in the diagram), unfettered and physically autonomous from the vast infrastructure that was the foundation of its lunar trajectory.

In Apollo's command or lunar module, by contrast, the astronaut's body would be literally hard-wired, connected to sensors reporting pulse, temperature, and breathing to the desk of a Mission Control technician a quarter-million miles away. (As we saw in Layer 18, it took the unprecedented crisis of Apollo 13 to break this long chain of telemetry.) The conception of the human body as a series of inputs, outputs, and controls is in turn representative of the cybernetic mindset of Apollo's systems managers.

Conceived by Norbert Wiener to describe the phenomena of feedback, homeostasis, and control common to natural and manmade systems, the cybernetic lens was cast in wartime, and would describe and control some of the most complex engines of destruction made by man. The great insight of Bernard Schriever in 1954, after all, was to see the Atlas ICBM not as an *object* at all (as had the project's previous managers at Convair), but rather as a set of relationships, a complex *system* of informational and material exchange. To manage the construction of such a system, Schriever created an organization in the system's own image, a complex network of bureaucratic feedback and material flows, wired into and out of the very same control rooms that would ultimately direct the weapon's internal controls to its target.

Once cast, however, the cybernetic lens of systems management was used to peer not just beyond the body but reflecting as well as refracting back into it. As military-industrial engineers sought to replace the warheads of systems-designed ICBMs with astronauts, the most literal result of this view was the 1958 cyborg proposal by Clynes and Kline, that the human body should itself be managed cybernetically for space travel just as the missile's collection of valves, circuits, and cowlings was. This was a literal as well as lexigraphical conflation of cybernetic controls and the human organism.

URBANE TISSUE

The second image provides a marked contrast to Clynes and Kline's cyborg. Instead of a web of systematic relationships, we are treated to a specular field, a slice of human body and ILC spacesuit

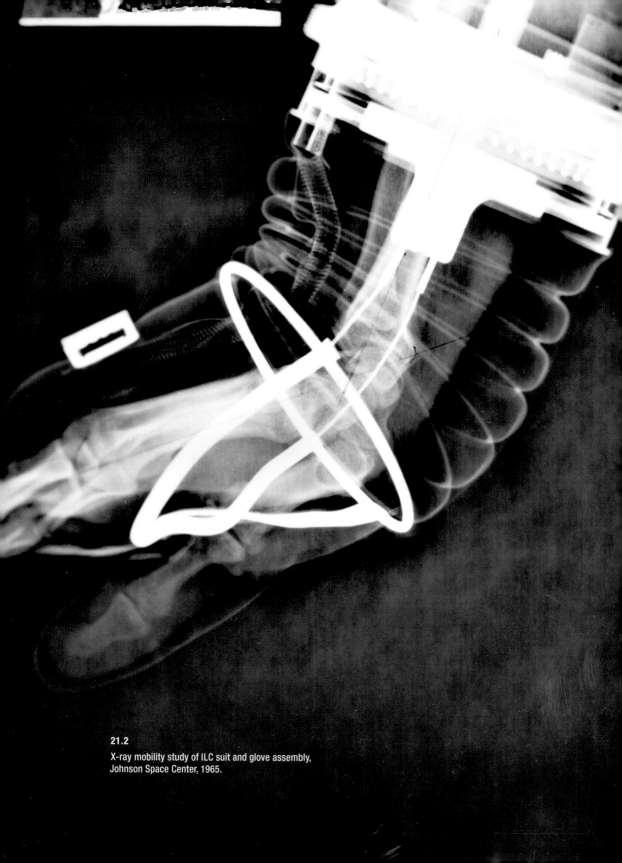

21.2
X-ray mobility study of ILC suit and glove assembly,
Johnson Space Center, 1965.

brought into focus on the surface of an X-ray. Here too there is an equivalence of natural and manmade, but (as is the nature of X-rays) the result is more than skin deep.

A ghost captured by the machine, the layered assemblage of the ILC spacesuit serves as both a mechanism to protect the body, and also a recapitulation of the body's own structure. Encapsulating the skeleton, it abandons a systems diagram in favor of the robust, epidermal logic of the body it contains.

In 1969, the systems engineer Simon Ramo published *Cure for Chaos: Fresh Solutions to Social Problems through the Systems Approach*.[1] In a now-familiar comparison, the book advocated the application of cybernetic techniques to the perceived disorder of its time: "If we can give an astronaut good air to breathe on the moon, then why not in our cities?"[2]

"[C]ities," Ramo continues, "do not constitute a good systems design. … Redesign to make them into a better overall system is not taking place at a sufficiently high rate."[3] As we regard Ramo's comparison between systems organization and urban "chaos," we should remember that, since Ramo's time, we have begun to see chaos itself as a complex kind of order. Deep in the tissue of the body itself, even in the ghostly marrow of the X-ray, interdependent networks eschew a logic of control for an interdependent logic of adaptation. Or, more properly, exaptation (see Layer 20). Just as the human skeleton is an extended system for phosphate storage, so is the ILC spacesuit an extended bra, or girdle. Its new function, shaped in a demanding network of adaptations, still carries the trace of an improbable origin.

This network of use and reuse turns us away from the pure logic of technology, and toward one of tissue—the fragile, layered membranes that are life's greatest strength. Here *incorporation* within a complex collaboration supersedes *invention* within a system of ideas and first principles: Something is made out of something else, both in terms of material, but also structure; a space suit is made out of a flight suit. An Apollo suit is made out of a Gemini suit, and also a Mercury suit, a pressurized flight suit, a Goodrich car tire, a bra, a girdle, a raincoat, a tomato worm. An American rocket ship is made out of a nuclear weapon, and a German ballistic missile; a "space program"—a new organization with new goals—is made out of preexisting military, scholarly, and industrial institutions and techniques. Within this process of continual adaption, there exist constructed and willfully iconographic landmarks, often built around images and visions. For all their novelty, however, such new looks are but momentary glances upon a state of constant appropriation and transformation.

FLESH AND FASHION

An uncharacteristic image of astronaut John Young points further. Uncharacteristic, because, as he poses in a David Clark Gemini spacesuit in 1963, he abandons the heroic, masculine posture rhapsodized by Barthes as the natural aspect of the "jet-man." Instead, he vamps. And the humor, of course, is in the incongruity—the masculinity of the fighter jock, the femininity of the pose.

But there is more here; a glance into the background of the image reveals a further, organizational irony that resembles and reflects the humor of the pose. Arrayed behind Young is a collection of custom-fitted Mercury acceleration couches, including his own. As we saw in Layer 12, it was only the provision of these custom-fitted chairs that allowed the systems design of the spacecraft to come within the Atlas missile's acceleration parameters.

21.3
Astronaut John Young posing in his David Clark
Gemini spacesuit, 1963.

Recapitulating the later logic of the layered A7L, the men of Mercury were effectively mated to the Ur-system of military-industrial production not through cybernetic control, but couture customization. Unlike any other part of the Mercury spacecraft, the couches were literally and figuratively "fashioned," custom-fitted to each user, and draped across their body in hard-setting plaster bandage before being molded in fiberglass.

In some settings—such as Mercury's custom couches—this juxtaposition between first principles and fashioning, the soft logic of the body and the hard logic of the system, can be seen as collaboration. But it was also, fundamentally, a conflict.

EXTENSION AND EXHAUSTION

This basic conflict is especially visible in a June 21, 1969, photograph of Neil Armstrong. Taken minutes after the conclusion of Apollo 11's EVA, it shows Armstrong returned to the (relative) safety of the lunar module, having unfastened his spacesuit helmet.

Visibly elated, Armstrong is also, clearly, exhausted. In a manner that would especially plague later Skylab astronauts (see Layer 18), the systems and schedules that transported him to the moon's surface did so in spite of, and not in sympathy with, the logic of his own body. Armstrong is shown radically extended, not in the cybernetic sense of augmentation, but in the literal sense of physical distance, and physiological exhaustion. Set against the control switches and visual austerity of the *Eagle*'s interior is the essential conflict between electronic order and robust, intimate disorder that defines the special softness of spacesuit, and spaceman.

A different picture of spacesuit and spaceman, NASA image AS11-40-5903, is one of the most widely reproduced images of the last century (and figure 1.1 of this book). Its chief subject, however, the ILC A7L, has been surprisingly absent from any design history related to man and space. Instead, curators and critics have focused on prototypes that are often literally, and always conceptually, much harder. Whether Litton's RX-5, Ames's AX-3, or even the AirResearch Advanced Apollo Prototypes,[4] we, perversely, privilege the failures of fabricating clothing for space, and not the successes.

Then again, we often fail to see that which is most apparent. In the case of the Apollo space-suit, both its Playtex origins and its epidermal structure seem to strike at a core of corporeal intimacy that we are embarrassed to address. This intimate, layered reality hides essential, and sometimes difficult, truths. We ignore them at our peril.

First among these is the continuing failure of systems thinking, however complex, to master the robust realities of human life, at any scale. Presenting a study of defense intellectuals' introduction of systems management to midcentury American planning, historian Jennifer Light asks an incisive question: "How and why are resources allocated time and again to support the adoption of technical and technological tools whose benefits remain unproven?"[5] This is in essence the same question we are asking here—why has the complex, layered reality of the Apollo spacesuit evaded any level of design inquiry? Why, instead, are unproven (if elegant) prototypes elevated as exemplary of good design?

The answers to questions about both design for the city on earth, and the body in space, are intimately related.

21.4

Neil Armstrong shortly after reentering the Lunar
Excursion Module *Eagle*, July 21, 1969.

CITIES AND FASHION

Let us examine once more the first of our new looks, Christian Dior's. A 1951 *Harper's Bazaar* photograph shows the model Mary Jane Russell wearing one of the most extensive of New Look dresses in front of Turgot's 1739 map of Paris. Prepared prior to the radical surgery of Haussmann, the Turgot map is a particularly apt backdrop to the complex, couture construction.

The New Look dress, of course, exerts a certain discomfort on the body of its wearer, shaping and girdling it to fit the fashion needs of its own day. Yet, in its multilayered construction, it is also deeply sympathetic to the skin underneath its folds. In this sense, it is a physical as well as conceptual parallel to the Apollo A7L, which we can consider in another, less familiar image, on the body of Apollo 12's Alan Bean. In his unfinished *Arcades Project* of 1927–1940, the German critic Walter Benjamin produced a complex mapping of Paris in the nineteenth century, in an interlinked selection of crafted and borrowed texts. Benjamin's word for these linked and layered texts, "convolutes," recalls, conceptually as well as topologically, the fashioned layers considered here, a quarter-million miles from the French capital.[6]

In 1962, John F. Kennedy's intimate understanding of the gulf between the body and its observed surface led to a particular sympathy for, and vision of, the importance of space exploration, and in particular the image of an American astronaut on the moon. Be it a picture of Camelot, Dior, or Apollo, we are not always willing, or able, to seek meaning under the surface that fashioned images represent. When we do, we discover that these new looks are, in fact, complex reassemblies of the old. Yet this is not mere trickery; the oldness of the new is itself a special kind of novelty.

The complex landscapes of cities, or the near-infinite biological web of the body, are both characterized by the same quality of "organized complexity" isolated by Warren Weaver in 1958. Subsequent study has revealed not only deep parallels between the complexity of cities and biological bodies,[7] but also increasingly emphasized that "there are strict limits to what can be predicted"[8] in the structure and behavior of each, complex assemblage.

In this sense, design for the body, as for the city, must take a different kind of lesson from nature than mimicking only its superficial shapes. With the notable exception of the Bauhaus costumiers who influenced spacesuit designer William Elkins (see Layer 17), it is normal for a fashion designer to fashion; that is, to start carefully from the known to suit the body to new circumstance, however radical that situation might be. In his 1940 essay "On the Concept of History," Walter Benjamin crafts a picture of fashion and urbanity that is, again, germane: "Fashion," Benjamin reflects, "has a nose for the topical, no matter where it stirs in the thickets of long ago; it is the tiger's leap into the past."[9] The image—*Tigersprung* in the German—fittingly captures the ceaseless appropriations and transformations that weave the fabric of urban life. It is, as ever, a timely lesson.

THE EARTH, FULL AND WHOLE

If an image rivals AS11-40-5903 for its visual proliferation in our culture, it is AS17-148-22727, the "blue marble," or whole earth.[10] Yet, as with the Apollo spacesuit's star turn, it is perhaps the intense reproduction of the image that hardens its surface to our view, and helps obscure essential lessons of its content.

We are often told of the inspirational role of this image in the environmental movement on earth, of the inconvenient truth represented by the borderless, interconnected globe. In fact, the image was as much a product of the green movement as a cause of it.[11] Apollo spacecraft had captured images of the earth's disk since 1968, but, given the particular requirements of lunar exploration, these were necessarily of the shadowed disk of our own planet, obscured by the darkness of space. The need to show relief in the lunar surface on landing, as well as to land on that portion of the moon's surface facing the earth (to allow for communication), meant that the sun was by necessity oblique to both moon and earth during all Apollo missions, showing both moon and earth half-full to inhabitants of either.

This changed in December of 1972, more than three years after Senator Gaylord Nelson first proposed a nationwide teach-in on ecological relationships, and two years after John McConnell's flag, with a painted full disk of the planet, had become the symbol for the resulting Earth Day in April 1970 (Senator Nelson had proposed a bill supporting the digital analysis of cities in 1965; see Layer 19). Responding to critiques of the space program's relevance to an increasingly distressed situation on earth (as well as hoping for the publicity that quickly evaporated from later Apollo efforts), NASA approved several firsts for the last Apollo mission. These included the first nonmilitary officer to be sent into space (geologist Harrison Schmitt), the first night launching of a Saturn V rocket, and, crucially, a modified launch trajectory that would allow an unshadowed, "full" earth to be photographed for the first time.

The wholeness of the image is familiar to us, but the particular quality of fragility it reveals is harder to see. As massive as our planet is (some 6 sextillion tons), the habitable environment of the globe, as revealed by our own fragility, is but a tissue-thin atmosphere.

The inextricably interrelated quality of this atmospheric space, as well as the increasing urbanization that shapes it, are the inescapable issues of our own age. When Buckminster Fuller proclaimed a "spaceship earth" in 1969, he evoked the nature of the earth, like a space capsule or the lunar spacesuit, as a closed system in the celestial void.[12] In the technological evangelism that characterized his enthusiasm, however, Fuller also sought to imply that, like the PLSS, or Saturn V, or the Atlas missile for that matter, the earth's closed system could be contained and controlled by a single diagram, given only a piece of paper large enough. His late-career efforts to create a "world resources inventory" and vast "Geoscope" at Southern Illinois University can be understood as a literal attempt to create a ground large enough for this phantom figure to appear upon.[13]

Following Fuller, we must indeed understand our world as similar to a spaceship; as in any closed system, we cannot escape the consequences of our own actions. But, equally, we must recognize that the robust complexity that characterizes our own body, cities, and planet will by its very nature subvert any attempt at diagramming or manipulating our environments with the precision of a spaceship's knob or dials.

As we face the necessity of transforming our own relationship to our only enduring spaceship, such a lesson is particularly essential. As the impact of human civilization on our robust yet fragile planet becomes ever more apparent, we must resist the seeming simplicity of the systems solution (as in recent proposals to "solve" global warming through systematic and global "geoengineering").[14] Much as the ILC A7L created a suitable space for man in the extreme landscape of the moon, so we must carefully—if sometimes radically—fashion a space for ourselves on earth.

21.5

A 1951 New Look ball gown modeled by Mary Jane Russell in front of Turgot's 1739 map of Paris, *Harper's Bazaar* (U.S. edition), October 1951.

21.6

Apollo 17 astronaut Harrison "Jack" Schmitt, standing by the Lunar Rover with the earth above; note that, as for all Apollo moon landing missions, the earth was in its first or last quarter as seen from the moon. Schmitt, a geologist, was the only nonmilitary astronaut to land on the surface of the moon.

21.7

The full earth photographed by Apollo 17 commander Eugene Cernan, December 1972. The image is shown here in its original orientation as photographed from the Apollo 17 Command Module *America*.

CONCLUSION: APOLLO'S GAZE

It has been the aim of this book to argue that the softness of the Apollo spacesuit is indicative of a special affinity between the bodies of Apollo's astronauts and the spacesuits that protected them—especially as differentiated from the remainder of Apollo's vast systems infrastructure. As such, their iconic image should project not a mastery of nature through technology, but rather a necessary sympathy to those parts of nature that, like our own bodies, defy easy systemization. Twenty-five years after walking on the surface of the moon, the normally reticent Neil Armstrong wrote his own letter of appreciation to the A7L suit and its makers. The suit, he wrote, "turned out to be one of the most widely photographed spacecraft in history. That was no doubt due to the fact that it was so photogenic. Its true beauty, however, was that it worked. It was tough, reliable, and almost cuddly."[15] The last but possibly best argument for this assertion is a matter not of poetry, but of history: the continued attention of the astronauts to their spacesuits, above all other surviving artifacts of the space age.

When in Washington, Apollo's elite alumni are frequent visitors to the Smithsonian Institution—but not to the vast Air and Space Museum that is the country's cathedral to space-age mythology. Rather, they travel regularly to an anonymous warehouse in Maryland, where the surviving Apollo spacesuits rest in storage, on bierlike shelves. The suburb is, quite coincidentally, called Suitland.

Unlike the rest of Apollo's hardware, it is clear, Apollo astronauts regard the suits as an extension of their own selves, and not a vehicle or container for them. Indeed, much like bodies—but unlike traditional "hardware"—the suits continue to change shape, size, and material, as their living, latex surfaces continue to adjust to the earth's atmosphere. Even as it represents a literal extension of a single Apollo astronaut's body, however, it is the lifelike quality of the A7L spacesuit—its physical material and robust, natural organization— that makes it represent all of our bodies as well. As we gaze upon the fabric spacesuit, all of us can look on the world with Apollo's gaze.

Notes

1.   Simon Ramo, *Cure for Chaos: Fresh Solutions to Social Problems through the Systems Approach* (New York: D. McKay, 1969).

2.   Ibid., 9.

3.   Ibid., 36.

4.   See Matilda McQuaid and Philip Beesley, *Extreme Textiles: Designing for High Performance* (New York: Smithsonian Cooper-Hewitt, National Design Museum, 2005).

5.   Jennifer S. Light, *From Warfare to Welfare: Defense Intellectuals and Urban Problems in Cold War America* (Baltimore: Johns Hopkins University Press, 2003), vii.

6.   See Walter Benjamin, *The Arcades Project*, ed. Rolf Tiedemann, trans. Howard Eiland and Kevin McLaughlin (Cambridge, MA: Belknap Press of Harvard University Press, 1999).

7.   See for example L. M. A. Bettencourt, J. Lobo, D. Helbing, C. Kühnert, and G. B. West, "Growth, Innovation, and the Pace of Life in Cities," *Proceedings of the National Academy of Sciences* 104 (2007): 7301–7306.

8. James Crutchfield, "The Hidden Fragility of Complex Systems—Consequences of Change, Changing Consequences," paper delivered to the symposium "Cultures of Change | Changing Cultures" (Barcelona, December 2009).

9. Walter Benjamin, "The Concept of History," in *Selected Writings*, ed. Howard Eiland and Michael William Jennings, vol. 4 (Cambridge, MA: Belknap Press of Harvard University Press, 2003), 395. For an extended discussion of the image of the *Tigersprung*, see Ulrich Lehmann, *Tigersprung: Fashion in Modernity* (Cambridge, MA: MIT Press, 2000).

10. See Denis Cosgrove, "Contested Global Visions: One-World, Whole-Earth, and the Apollo Space Photographs," *Annals of the Association of American Geographers* 84, no. 2 (1994): 270–294.

11. As but one example, see Felicity Scott, *Architecture or Techno-Utopia: Politics after Modernism* (Cambridge, MA: MIT Press, 2007). The last plate of this otherwise virtuoso reading of late-1960s/ early-1970s architecture is AS17-148-22727, misidentified as being taken from Apollo 11 in 1969.

12. See R. Buckminster Fuller, *Operating Manual for Spaceship Earth* (Carbondale: Southern Illinois University Press, 1969).

13. Lloyd Steven Sieden, *Buckminster Fuller's Universe: His Life and Work* (Cambridge, MA: Perseus Publishing, 1989), 259–269.

14. See James Rodger Fleming, "The Pathological History of Weather and Climate Modification: Three Cycles of Promise and Hype," *Historical Studies in the Physical and Biological Sciences* 37, no. 1 (2006): 3–25.

14. See James Rodger Fleming, "The Pathological History of Weather and Climate Modification: Three Cycles of Promise and Hype," *Historical Studies in the Physical and Biological Sciences* 37, no. 1 (2006): 3–25.

15. Neil Armstrong, letter to "EMU Gang, Johnson Space Center," July 14, 1994. Courtesy Crew Systems Division, Johnson Space Center, Houston, TX.

Illustration Credits

1.1   NASA Image AS11-40-5903, courtesy Johnson Space Center.

1.2   NASA Photo 69-HC-718, courtesy Kennedy Space Center.

1.3   NASA Image 61-MR3-109, Record Group 255-G-94, Still Picture and Film Collection, National Archives at College Park, College Park, MD.

1.4   NASA Image S69-55368, courtesy Johnson Space Center.

2.1   Engraving from the collection of the Army Air Forces, Record Group 342 AAF, Image BN9071, National Archives at College Park, College Park, MD.

2.2   Image from the *Illustrated London News*.

2.3   Record Group 342 AAF, courtesy National Archives at College Park, College Park, MD.

3.1   Photograph by Willy Maywald, 1955, © 2008 Artists Rights Society (ARS), New York / ADAGP, Paris, image courtesy V&A Images, Victoria and Albert Museum, London.

3.2   Photograph by Loomis Dean, *Time-Life*, image courtesy Getty Images.

3.3   Photograph by Henri Cartier-Bresson, *Magnum*, image courtesy Magnum Photos.

4.1   Air Force photograph 45903 AC, Record Group 342 FH, Still Picture Collection, National Archives at College Park, College Park, MD.

4.2   Image B44539, Record Group 342 FH, Still Image and Photo Collection, National Archives at College Park, College Park, MD.

4.3   Image 26-15057-62, Record Group 347 ANT, National Archives at College Park, College Park, MD.

4.4   U.S. Air Force Image K22386, Record Group 342B, National Archives at College Park, College Park, MD.

4.5   U.S. Air Force Image 167636, Record Group 342B, National Archives at College Park, College Park, MD.

5.1   © 1935, renewed 1962 Columbia Pictures Industries Inc., courtesy Columbia Pictures.

5.2   Record Group 342 FH, Still Picture Collection, National Archives at College Park, College Park, MD.

5.3   Record Group 342 FH, Still Picture Collection, National Archives at College Park, College Park, MD.

5.4   Record Group 342 FH, Still Picture Collection, National Archives at College Park, College Park, MD.

6.1 Image from Getty Images, first published in *Life* magazine.

6.2 Courtesy Rockland State Hospital/Nathan Kline Institute Archives.

6.3 Illustration by Fred Freeman. Every attempt has been made to contact the artist's estate. If you have further copyright information, please contact the author.

6.4 Cartoon from *Astronautics*, March 1960, 31.

7.1 Image SI-2004-19125, courtesy Smithsonian Institution, Washington, DC.

7.2 *New York Times* (March 30, 1941), courtesy David Clark Company, Inc.

7.3 Image 342FH-3B-28525, courtesy National Archives at College Park, College Park, MD.

7.4 Image 342-FH-3B-34751, courtesy National Archives at College Park, College Park, MD.

7.5 Images 342-FH-3B-37457a and 342-FH-3B-37458a, courtesy National Archives at College Park, College Park, MD.

7.6 © Hergé/Moulinsart 2008.

7.7 Image 342-FH-3B-37460, courtesy National Archives at College Park, College Park, MD.

7.8 Image 342-B-03-13-14-01, courtesy National Archives at College Park, College Park, MD.

7.9 NASA Image S63-07852, courtesy Johnson Space Center.

7.10 NASA Image S65-30433, courtesy Johnson Space Center.

7.11 NASA Image S66-38082, courtesy Johnson Space Center.

8.1 Collection of the author.

8.2 Smithsonian Image NASM-7B34428, courtesy National Air and Space Museum.

8.3 Smithsonian Image SI-88-14980, courtesy National Air and Space Museum.

8.4 Collection of the author.

9.1 *Life*, April 1, 1940, 7, courtesy ILC Dover, Inc.

9.2 Courtesy ILC Dover, Inc.

9.3 Courtesy ILC Dover, Inc.

9.4 NASA Image AS12-49-7278, courtesy Johnson Space Center.

10.1 © 2008 Artists Rights Society (ARS), New York/DACS, London as well as Tate, London/Art Resource, NY.

10.2 AP Image, ID 610621014.

10.3 Image © Associated Press, courtesy John F. Kennedy Library, Boston.

10.4 AP Photo by Malcolm Browne, ID 630611023.

10.5 AP Image, ID 6305030121.

10.6 AP Image, ID 6202200132.

10.7 Image © The Herb Block Foundation.

10.8 Photo 62-ADM-13, courtesy NASA.

11.1 NASA Image B-60-1496, courtesy Glenn Research Center.

11.2 U.S. Air Force Image 342-AH-B44810, courtesy National Archives at College Park, College Park, MD.

11.3 U.S. Air Force Image 98922, Record Group 342-AH, courtesy National Archives at College Park, College Park, MD.

11.4 U.S. Air Force Photograph 342-B-04-070-10-16250AC, National Archives at College Park, College Park, MD.

11.5 NASA Image C1960-53088.

11.6 NASA Image C1959-52233, courtesy Glenn Research Center.

12.1 NACA Image A-20419, Record Group 255 G Ames 11, Still Photo Collection, courtesy National Archives at College Park, College Park, MD.

12.2 Image 342-B-03-014-12-157268AC, courtesy National Archives at College Park, College Park, MD.

12.3 Image 342-B-03-014-12-157241AC, courtesy National Archives at College Park, College Park, MD.

12.4 NASA Image LARC-M-303, courtesy Johnson Space Center.

12.5 NASA Image 65-H-703, courtesy Johnson Space Center.

12.6 Image 255-G-38-63-Gemini-47, courtesy National Archives at College Park, College Park, MD.

12.7 NASA Image S-67-34085, courtesy Johnson Space Center.

12.8 NASA, Box 33, courtesy JSC History Collection, Center Series, Center Flight Crew Operations Collection, University of Houston-Clear Lake.

12.9 Dryden Flight Research Center Image # ECN-506, courtesy Dryden Flight Research Center, Edwards Air Force Base, Boron, CA.

12.10 Image 255-CB-64-H-2459, courtesy National Archives at College Park, College Park, MD.

12.11 Collection of the author.

13.1 NASA Image 63-Apollo-106, Record Group 255-GC, Still Photo Collection, courtesy National Archives at College Park, College Park, MD.

13.2 Photograph by Mark Avino, photograph 2003-27317, courtesy Smithsonian Institution.

13.3 NASA Image S-63-16608, courtesy JSC History Collection, Center Series, Spacesuit Collection, University of Houston-Clear Lake.

13.4 NASA Image S-64-23429, July 21, 1964, Project Control Files, National Aeronautics and Space Administration, Manned Spacecraft Center, Engineering and Development Directorate, Crew Systems Division, Record Group 255-E171B, National Archives and Records Administration—Southwest Region (Fort Worth, TX).

13.5 NASA Image 65-H-208, Record Group 255-CB, Still Photo Collection, courtesy National Archives at College Park, College Park, MD.

13.6 NASA Image S65-36332, courtesy JSC History Collection, Center Series, Spacesuit Collection, University of Houston-Clear Lake.

13.7 Photograph by Mark Avino, Smithsonian Image 2003-27279, courtesy Smithsonian Institution, National Air and Space Museum.

13.8 Photograph by Mark Avino, Smithsonian Image 2003-27277, courtesy Smithsonian Institution, National Air and Space Museum.

13.9 Captured frames from NASA copy of film, courtesy Johnson Space Center.

13.10 NASA Image AS17-143-21941, courtesy Johnson Space Center.

14.1 Courtesy ILC Dover, Inc.

14.2 Courtesy ILC Dover, Inc.

14.3 Courtesy ILC Dover, Inc.

14.4 Courtesy ILC Dover, Inc.

14.5 Courtesy ILC Dover, Inc.

14.6 Courtesy ILC Dover, Inc.

14.7 Courtesy ILC Dover, Inc.

14.8 Courtesy ILC Dover, Inc.

14.9 Courtesy National Gallery of Australia and VAGA, Inc.

14.10 Courtesy ILC Dover, Inc.

14.11 Courtesy ILC Dover, Inc.

14.12 Courtesy ILC Dover, Inc.

15.1 File Litton Mark I SI-86-14174.tif, Image SI-86-14174, courtesy William Elkins.

15.2 Courtesy San Francisco Public Library.

15.3 NASA Image 64-Apollo-104, Record Group 255-G, Still Picture and Photograph Collection, National Archives at College Park, College Park, MD.

15.4 NASA Photo S82-37401, courtesy Johnson Space Center.

15.5 NASA Photo S-65-57409, courtesy Johnson Space Center.

15.6 Project Control Files, National Aeronautics and Space Administration, Manned Spacecraft Center, Engineering and Development Directorate, Crew Systems Division, Record Group 255-E171B, National Archives and Records Administration—Southwest Region (Fort Worth, TX).

15.7 Project Control Files, National Aeronautics and Space Administration, Manned Spacecraft Center, Engineering and Development Directorate, Crew Systems Division, Record Group 255-E171D, National Archives and Records Administration—Southwest Region (Fort Worth, TX).

15.8 NASA Image S-68-40473, courtesy Johnson Space Center.

15.9 Record Group 306HVM, National Archives Still Picture and Photograph Collection, National Archives at College Park, College Park, MD.

15.10 Courtesy Davis Brody Bond LLC.

15.11 Courtesy Davis Brody Bond LLC.

16.1   Courtesy NASA.

16.2   Smithsonian Image SI-2000-10046, Joel Banow Collection. © CBS.

16.3   Smithsonian Image SI-2000-10052, Joel Banow Collection. © CBS.

16.4   Smithsonian Image SI-2000-10049. © CBS.

16.5   Courtesy Library of American Broadcasting, University of Maryland Libraries.

17.1   Courtesy NASA Ames Research Center and Vic Vykukal.

17.2   Courtesy NASA Ames Research Center and Vic Vykukal.

17.3   NASA Image S82-2644, courtesy Johnson Space Center.

17.4   Courtesy NASA Ames Research Center and Vic Vykukal.

17.5   NASA Photo JSC S88-28962, courtesy Johnson Space Center.

18.1   Images 342b-05-010-1-153241AC and RG 342b-05-010-1-150121AC, declassified February 1955, National Archives at College Park, College Park, MD.

18.2   Image RG 342b-11-082-2164994AC, National Archives at College Park, College Park, MD.

18.3   Image 342b-11-082-1-168810, National Archives at College Park, College Park, MD.

18.4   Image 255-GC-MA8-20a, National Archives at College Park, College Park, MD.

18.5   NASA Image S63-22315, courtesy Johnson Space Center History Office.

18.6   NASA Image S66-24806, courtesy Johnson Space Center.

18.7   Image 255-CB-64-H-2346, Record Group 255, National Archives at College Park, College Park, MD.

18.8   Image 342B-11-061-1/5-172310AC, Record Group 342-B, National Archives at College Park, College Park, MD.

18.9   NASA Image S65-13854, courtesy Johnson Space Center.

18.10   NASA Image S69-39815, courtesy Johnson Space Center.

18.11   NASA Image AP11-69-H-1121, courtesy Kennedy Space Center.

18.12   NASA Image AS13-62-8929.

18.13   NASA Image S74-17304, courtesy Johnson Space Center.

18.14   NASA Image 63-Apollo-155, courtesy Johnson Space Center.

19.1   NASA Image C-52113, courtesy NASA Lewis Research Center.

19.2   UCLA Urban Laboratory, 1967.

19.3   Image 207_MPF-199-22, Record Group 207, National Archives at College Park, College Park, MD.

19.4   Richard D. Johnson and Charles Holbrow, eds., *Space Settlements: A Design Study*, NASA SP-413 (Washington, D.C.: Scientific and Technical Information Office, National Aeronautics and Space Administration, 1977), 49.

20.1   Courtesy ILC Dover, Inc.

20.2   NASA Image S-65-4973, courtesy Johnson Space Center.

20.3   NASA Image S-65-4970, courtesy Johnson Space Center.

20.4   NASA Image S-69-2649, courtesy Johnson Space Center.

20.5   Courtesy ILC Dover, Inc.

21.1   Project Control Files, National Aeronautics and Space Administration, Manned Spacecraft Center, Engineering and Development Directorate, Crew Systems Division, Record Group 255-E171B, National Archives and Records Administration—Southwest Region (Fort Worth, TX).

21.2   Center Spacesuit Series, Box 9, courtesy Johnson Space Center Archive, University of Houston Clear Lake.

21.3   NASA Image S-63-15077, courtesy Johnson Space Center.

21.4   NASA Image AS11-37-5528, courtesy Johnson Space Center.

21.5   Photograph by Louise Dahl-Wolfe, courtesy Staley-Wise Gallery, New York.

21.6   NASA Image AS17-134-20471, courtesy Johnson Space Center.

21.7   NASA Image AS17-148-22727.

Index

Page numbers in *italics* indicate illustrations.

Insulation, 127, 238
Intelligence, 177, 228
International Business Machines (IBM) Corporation,
    72, 168, 172, 175, 177, 273, 293, 296n
  360 mainframe computer, 175, 279, 283
  4Pi/AP-101 computer, 293, 296n
  7030 "Stretch" mainframe computer, 282–283
  7090II mainframe computer, 283
  Houston Automated Spooling Priority (HASP),
    283
  Mission Control and, 175, 277–279, 282
International Latex Corporation (ILC), 6, 7
  advertising and, 117, 121
  Air Force and, 92, 94–95, 124
  AirResearch and, 264
  Ames AX suits and, 263–264, 267
  corporate history, 119, 124, 128
  Hamilton Standard and, 7, 126, 182–183, 189,
    191–194, 199, 214
  latex convolute, 124, 183, 209
  latex manufacturing and, 117
  Ling-Temco-Vought (LTV) and, 215
  Litton Industries and, 228, 232, 238, 241–242
  Playtex, consumer brand, 3, 6, 7, 8, 33, 92, 116,
    118–123, 127, 179, 182, 198–199, 209,
    211, 228, 293, 333
    brassieres, 211, 311
    girdles, 33, 117, 311
  sewing techniques, 209, 211, 219
  spacesuit design and, 126–127, 182–183, 189,
    191–194, 198 (*see also* Apollo program:
    A7L spacesuit; Apollo program: A7LB
    spacesuit)
  SPD-143 prototype, 183, *184*, *185*
  systems engineering and, 191, 198–204, 209,
    211, 214–216, 219, 223–224, 270
  World War II and, 120–121
Invention, 15, 20, 31, 68–73, 118–119, 124, 159,
    179, 229–230, 300, 331
Iron Curtain, 88

Jacobs, Jane, 307–308
Jacobson, Max, 133, 137
Jaquard, Joseph-Marie, 16
Jazz, 20
Jet age, 65, 127
Johnson, Lyndon Baines, 40, 106n, 137, 140,
    142, 301

media relations, 251–252
NASA and, 143–144, 155
Johnston, Richard S., 126, 182, 191–192, 214–215
Joints, 111, 183, 191–192, 232–238, 263–267,
    319, 325

Kamnitzer, Peter, 175, 300
Kelly, Burnham, 298
Kennan, George F., 38
Kennedy, Jacqueline Bouvier, 133, 137
Kennedy, John Fitzgerald
  and Corona spy satellites, 143
  Hamilton and, 6, 132
  hats and, 138
  illness, 132–140
  lunar landing decision, 6, 143–147
  media coverage, 99, 140, *141*
  personal appearance and grooming, 138–139
Kennedy, Joseph P., 133, 137
Kennedy, Robert F., 133, 219
Khrushchev, Nikita, 109, 143
Kittinger, Joseph, 152–154, *157*, 159
Kline, Nathan, 6, 68–77, 204, 292, 299, 310,
    317–321, 324, 329
Korolev, Sergei, 109
Kosmo, Joe, 234
Kraepelin, Emil, 69
Kraus, Hans, 133
Kubrick, Stanley, ix
  Doctor Strangelove, 40
  2001, 7, 255, 303, 310
KV-2 Russian spacesuit, 111

Laboratory, 41, 70–73, 83, 85, 92, 127, 154, 230,
    234, 239, 299
  space laboratory, 228, 233
Lamanova, Nadezhda, 269
La Mettrie, Julien Offray de, 23
Laminates, 264
Landing, 18–20, 58, 64–65, 92, 95, 99, 109–111
  lunar, 143–144, 168, 172–176, 187, 198,
    234, 240–242, 251–259, 277, 281,
    292–293, 336
  splashdown, 104, 109, 292
Lasker, Mary, 68
Latex, 16, 111, 117, 119–121, 127, 164, 183, 189,
    193, 209–211, 253, 255, 267, 342

in materials, 317
in military attire, 90
in psychiatry, 71, 77n
News coverage, 20, 47, 65, 68, 109, 121, 132, 252, 257. *See also* Media; Press
Nixon, Richard M., 132, 251, 291, 303, 306–307
North American Aviation
    Apollo Command Module, 172, 182–183, 187–189, 214, 241–242, 262, 287, 293
    Saturn V rocket, 172, 182, 240, 255, 287, 337
    Space Shuttle, 95, 267, 293, 326n
    X-15 aircraft, 94–95, 99, 104, 124, 126
Nuclear development and defense, 37–54, 86–88, 159, 168, 273, 281, 303, 306, 331
Nylon, 65, 85, 94, 99, 124, 126, 183, 211, 215, 232, 317. *See also* DuPont

Operation Breakthrough, 303–307, *304*, 310
Orbit, 22, 47, 76, 104, 109, 112, 132, 143, 144, 149, 168, 239–242, 252, 255, 264, 267, 277–279, 291–292
Organization
    Cold War, 37
    defense, 45–51
    in fashion, 31
    institutional, 127, 159, 182–183, 191, 194, 198–199, 204, 211, 224, 252, 321–324
    organized complexity and, 308, 310, 336
    Soviet, 114
    systems, 329, 331
    technological, 279, 287

Packaging, 118–123, 317
Paine, Thomas O., 299, 303, 321, 323
Paley, Rev. William, 321
Paley, William (CBS president), 252
Partial-pressure suits. *See* Pressure suits
Patents, 82–83, 117, 120, 126, 164, 192, 229
Patterns, 16, 86, 111, 209, 214, 216, 219
Performance, 50
    aircraft, 81–82, 172, 263
    computer simulation, 167, 172, 175
    human, 57, 59, 151
    pressure suit components and, 82–86, 94, 126, 181–206, 241, 317
Permeability, 183
Philco-Ford, 277, 281, 292

Physiology, 23, 59, 64, 75, 85, 150–162, 204.
    See also Human body
Piccard, Auguste, 20
Pilâtre de Rozier, Jean-François, 16
Pilkenton, Roberta, 130n, 209
Pilots, 17–20, 24, 39, 41, 57–65, 72, 81–104, 109–111, 123, 152–155, 164–172, 191, 263, 273, 291
Playtex. *See* International Latex Corporation
Pleats, 92, 232, 234, 263
Pneumatics, 61, 64, 83, 119, 164, 199, 279, 305, 306
Polignac, Gabrielle de Polastron, duchesse de, 17
Population, 17, 251, 298, 305
Portable life support system (PLSS), 189, 198, *328*, 329, 337
Post, Wiley
    altitude record, 64
    circumnavigation records, 57, 58–59
    jet lag, discovery of, 59, 64
    jet stream and, 64
    media images, *56*, 57, *58*, 64
    pressure suit innovations, 6, *60*, 61, *62–63*, 89–90
Pouge, William, 289, *290*, 291–292
Precision management, 192
Press, 20, 29, 40, 57, 76, 86, 233, 255, 263, 303, 337. *See also* Media; News coverage; Publicity
    Kennedy and, 137–140
Pressure
    air, 19, 117, 151, 232, 324
    bladder, 61, 81, 104, 111, 127, 192–193, 219
    cabin or enclosure, 20, 23, 59, 64, 85–99, 109, 232–233
    low, 59, 61
    research lab, 92, 94
Pressure suits, 24, 57–65, 75, 79–106, 109, 111, 124–127, 132, 154, 181–206, 211, 219, 227–248, 331
Production
    aircraft, 94, 182
    fabric, 16
    flight/space suit, 85–86, 90, 99, 117, 182, 191, 204, 211–216, 230, 263
    high-technology, 191, 232, 317
    housing, 306

line, 121, 198, 209
mass, 119–121, 126, 159, 230, 303
underwear, 33, 82–83, 211
weapons, 40–46, 240, 301, 317
Program-planning budgeting system (PPBS), 321
Projection, 50, 171, 277–279
Project Manhigh, 152–154, *157*, 159
Project Whirlwind, 166–168, *169*, 176–177, 273
Propaganda, 143, 144
Prosthetics, 16, 72, 75, 83, 154, 269
Prototypes, 39, 40, 46, 50, 61, 83–104, 126, 159,
        166, 182–204, 209, 214, 234–247,
        263–264, 269–270, 303, 333
Psychiatry, 75
Public, 17, 22, 31–33, 47, 57, 86, 121, 143, 194
    demonstration, 16
    imagination, 23, 59
    perception, 71, 104, 137, 143, 159, 242, 251,
        267
    spectacle, 15, 140
Publicity, 29, 40, 57, 64, 109, 123, 132, 138, 307,
        337. *See also* Media; Press

Radford, Arthur W., 37, 39
Radio, 230
    broadcast, 111, 229
    modern, 20
    signal, 20, 232, 283
Radnofsky, Matt, 194, 269
Ramo, Simon
    systems engineering and, 45–47, 50, 159, 211,
        229, 301, 303, 321
    urban planning and, 331
RAND Corporation, 40, 299–301, 306
Rauschenberg, Robert, 216, 219, *220*, 223
Raytheon Company, 175
    Apollo Guidance Computer, 172, 282, 293–294
Reproduction, 15–16, 109, 223–224, 310, 333, 336
Research
    defense, 45, 83, 88, 301, 303
    flight physiology, 83, 151–152
    materials, 33, 83, 118–119, 317
    medical and mental health, 68–77, 154
    pressure suit, 89–95, 121–126, 183, 192
    rocket, 39–40
    space, 229–234, 264, 267, 292, 301, 310
Revolution. *See also* "New Look"
    fashion, 29, 82

flight, 81, 82
information, 73
psychiatric, 70–71
technological, 223
women's underwear, 117–121
Rheim, Homer, 191
Rivarol, Antoine de, 15
Rockland State Mental Hospital, 68–71, 73, 76, 77n
Romney, George, 303, 306
Roosevelt, Franklin, 24, 121, 164
Rubber, 58, 61, 83–92, 111, 117–121, 126–127,
        209, 211, 230–232, 281
    bladder, 92, 95, 209, 235
    girdle, 33, 117
Rudolph, Paul, 40
RX suits. *See* Litton Industries

Satellites, 111, 143, 232–233, 242, 267
Saturn V rocket, 172, 182, 240, 255, 287, 337
Schirra, Walter, 106n, 257
Schlemmer, Oskar, 230, 269
Schmitt, Harrison "Jack," 337, 339
Schriever, Bernard H.
    systems engineering and, 45–47, 50–51, 159,
        211, 277, 300, 321–323, 329
    urban design and, 301, 303, 306
Science and the City, 300, 308
Science fiction, 57, 65, 104, 154, 228
Scott, Dave, 168, 176
Scott-Heron, Gil, 259
Seamstresses, 85, 126, 204, 209, 211, 214, 224,
        325
Sewing, 33, 85, 111, 126, 191, 193, 209, 211,
        214–219, 224, 263, 269–270
Shepard, Alan, *9*, 11, 81, 144, 192, 291
Sheperd, Leonard "Lenny," 124, 126, 191–194, 211,
        215
Shockley, William, 232
Sidey, Hugh, 138
Signals, 72–73, 229. *See also* Radio: signal
    biological, 73
    electrical, 20, 165, 168
    television, 251–255
Simulation, 163–180, 281
    Cold War, 300
    computerized, 51, 166–168, 281–282
    digital images and, 175, 253–255, 257
    flight, 83, 86, 151, 193

Theater, 68, 273, 279, 281
Thermal Micrometeoroid Garment (TMG), 187, 209, 223
Thich Quang Duc, 140, *141*
Thompson, Wayne, 298, 310
Thompson Ramo Wooldridge. *See* TRW Inc.
Thornton, Charles "Tex," 45, 228–229, 232, 248n, 305
Tigersprung, 336
Tiger team, 191–192, 204
Time (magazine), 46, 50
Tintin, 90, *91*, 240
Tissandier, Gaston, 19
Tomato worm, 90, 124, 232, 331
Training, 112, 133, 137, 151, 159, 164, 166, 168, 171–172, 176, 183, 192, 193, 219, 228, 273, 293
Transistor, 73, 172, 177, 230, 232, 293
Travel, Janet, 133
Truman, Harry S., 37
TRW Inc.
    as Ramo-Wooldridge, 46–47, 50, 211, 303
    as Thompson Ramo Wooldridge, 211, 303
Turgot map of Paris, 336
Turnbull, Douglas, 7, 255
    "HAL 10000" CBS set, 253, 255, *256*
2001 (film), 7, 255, 303, 310

Underwear
    brassiere, 86, 117–118, 121, 211
    corset, 82–86, 90, 117–120, 137
    girdle, 29, 33, 82, 92, 117–127, 182, 198, 209, 211, 219, 319, 325, 331
    revolution in women's, 117–121
Uniforms, 46, 59, 104, 109, 121, 123, 127
Union of Soviet Socialist Republics. *See* Soviet Union
UNIVAC, 166–167
University of California, Berkeley, 298, 301
University of California, Los Angeles (UCLA), 175, 300
University of Southern California (USC), 85–86, 311n
Urbanism
    complexity, 300–305, 331
    crisis, 298–303, 307
    cybernetic model, 299
    form, 292, 299, 305

landscape, 308
planning, 307–308
redevelopment, 306, 307–308
suburban, 279, 281, 299
systems, 300–306
technology and, 299–307
Urban Systems Associates Inc. (USA), 301, 306
Urinary Collection Device, 223
Utopia, 15, 247, 269, 310

Vacuum, 75, 95, 99, 118–119, 124, 199, 204, 211, 228–232, 287
    suit, 230–233
    tube, 22, 73, 130, 166–167, 177, 228–232
Vail, Edwin, 191
Vaucanson, Jacques, 15–16
Vaudoyer, A. T. L., 16
Videogames, 177, 179
Vinyl, 85
Virtual
    aircraft, 167
    data, 73
    display, 171–172
    pilot, 165
    space, 164–168, 176–179, 273, 282, 292–293
Visual effects, 253, 255
Von Braun, Wernher, 12n, 39, 85, 144, 159, 240, 277
Von Kármán, Theodore, 22
Von Neumann, John, 40, 45, 167
Voshkod space capsule, 111–112, 114n
Vostok space capsule, 6, 109, *110*, 111–112
Vrba, Elisabeth S., 319
Vykukal, Hubert "Vic," 263–264, 269–270

Wagner, Andreas, 324–325
Weapons, 72–73. See also Atomic weapons; Nuclear development and defense
    arms race, 39, 46, 51, 81, 143, 168, 277
Weaver, Robert, 301
Weaver, Warren, 308, 310, 336
Webb, James, 12n, 143, 216, 253, 298–299, 303, 321
White, Edward, 99, *100–101*, *170*
Whole Earth, photograph, 336–337, *340*
Wickham, Henry, 119
Wiener, Norbert
    cybernetic theories, 72–73, 167, 329